Exploring Marine Biology

Laboratory and Field Exercises

Paul A. Haefner, Jr.

Rochester Institute of Technology

New York   Oxford
OXFORD UNIVERSITY PRESS

LAB MANUAL DISCLAIMER

Performance of lab experiments requires careful attention to safety, and much thought has been given to the safety of the students in the preparation of the exercises in this Laboratory Manual. All exercises contained in the manual have been performed by the author. When performing any lab exercise, it is imperative for students to follow all lab and safety instructions, both those presented in this manual and those given by the lab instructor. If performed with the materials and equipment specified in this text, in accordance with the methods developed in the text, the author believes the exercises to be a safe, valuable, educational experience. However, all duplication or performance of these exercises is conducted at one's own risk. The author and the publisher cannot warrant or guarantee the safety of individuals performing these exercises; and they hereby disclaim any liability for any loss or damage claimed to have resulted from or related in any way to the exercises, regardless of the form of action.

Oxford University Press

Oxford   New York
Athens Auckland Bangkok Bogotá Buenos Aires Calcutta
Cape Town Chennai Dar es Salaam Delhi Florence Hong Kong Istanbul
Karachi Kuala Lumpur Madrid Melbourne Mexico City Mumbai
Nairobi Paris São Paulo Shanghai Singapore Taipei Tokyo Toronto Warsaw

and associated companies in
Berlin Ibadan

Copyright © 2001 by Oxford University Press, Inc.

Published by Oxford University Press, Inc.
198 Madison Avenue, New York, New York, 10016
http://www.oup-usa.org

Oxford is a registered trademark of Oxford University Press

Library of Congress Cataloging-in-Publication Data available upon request

ISBN 978-0-19-514817-6

Printed in the United States of America
on acid-free paper

Dedication

To my parents, who started all this with trips to Rehoboth and Ocean City, and made it possible; the faculty at Franklin & Marshall, especially Mac Darlington, Art Shively, and Harry Lane, for their moral support and encouragement; John L. Atlee, Jr., M.D., who provided the financial support for the MBL (Woods Hole) opportunity; Carl Shuster and Frank Daiber, for giving a landlubber a graduate education at the Bayside Laboratory; and especially, Susie, my wife, who likes sandy beaches and sea breezes, and tolerated so much time away from them.

PREFACE

Exploring Marine Biology: Laboratory and Field Exercises is a manual suitable for all marine biology courses. It can be used both in the one-term introductory course in marine biology typically taken by students who have had an introductory course in the biological sciences, and in introductory courses designed for the nonscience major.

Each unit features a variety of exercises designed to allow instructors to select the desired extent and depth of coverage that best fits their approach, their students' background, and the time and material available. Suggestions for optional, independent studies are also included.

Flexibility of This Manual

Because many of the exercises emphasize the use of living organisms and collections available from marine supply houses, the exercises are applicable to courses taught at either coastal or "inland" institutions. The availability of a wide range of marine aquaria and supplies supports those courses dependent on the import of organisms.

In writing a manual such as this, it is somewhat difficult to make all exercises perfectly applicable to all the diverse marine habitats of North America; however, such an attempt has been made. Many of the examples and the three extensive, illustrated keys featured in this manual include representative organisms from a variety of marine habitats, not just a particular region.

Organization of This Manual

Unit 1 begins with an analysis of four basic parameters used to describe water quality. Unit 2 covers classification and identification of invertebrate taxa most often encountered by students throughout the course and supports the conspicuous domination of marine life by invertebrate animals (at least ninety percent of all described species are invertebrates). The unit features an illustrated key to macroscopic coastal marine invertebrates. Unit 3, on benthic animals, continues the invertebrate theme, with emphasis on the study of representative forms of different feeding types. This investigation is followed in Unit 4 by the microscopic study of meiofauna, the interstitial animals of ben-

thic sediments. Morphology and diversity of macroscopic algae and sea grasses are then presented in Unit 5, prior to the coverage of epibiota of algae and sea grasses in Unit 6. A study of photosynthetic pigments follows in Unit 7. The extensive coverage of plankton in Unit 8 focuses on phytoplankton and the importance of copepods (grazers) and arrow worms (predators) and includes identification and quantitative analysis of plankton samples. The unit also includes an illustrated key to plankton. Fishes are presented as representatives of nekton, with equal balance between recognizing fish morphology and diversity of form in Unit 9, which contains an illustrated key to orders of fishes, and functional anatomy in Unit 10. The manual then shifts emphasis from studies of individual organisms to qualitative and quantitative analyses of fouling communities in Unit 11. Field-oriented exercises include sampling intertidal habitats of the seashore in Unit 12 and deeper offshore habitats by research vessel in Unit 13.

Three appendices include instructions for (1) the use of microscopes and ocular micrometers, (2) data analysis, and (3) writing laboratory and field trip reports.

Instructor's Guide

The *Instructor's Guide* for *Exploring Marine Biology: Laboratory and Field Exercises* includes information on safety in the lab; a list of resources for materials, organisms, videos, and field guides; a list of laboratory materials and reagents common to all units; specific suggestions for setting up and performing each laboratory unit; and examples of possible field projects.

Acknowledgments

I am indebted to the editors of D. C. Heath who encouraged this endeavor and provided unfailing support throughout the planning, writing, reviewing, and revising process. And to the following reviewers, who provided constructive critiques of the manuscript, a special note of thanks: D. V. Aldrich, Texas A&M–Galveston; Jane Aloi, Saddleback College; James P. Beets, University of Richmond; Mark C. Benfield, Woods Hole

Oceanographic Institution; Tammy S. Brannon, Allan Hancock College; Paula F. Dehn, Canisius College; Harry R. Greer, Cayuga Community College; Andrea L. Huvard, California Lutheran University; John T. Lehman, University of Michigan; Dewey H. Lewis, Coastal Carolina Community College; Mary Sue Lowery, University of San Diego; Lisa K. Muehlstein, University of Richmond; Gail L. Ogden, Moorpark College; Paulette Peckol, Smith College; Michael C. Smiles, Jr., SUNY–Farmingdale; Bruce A. Thompson, Louisiana State University.

P. A. H.

Purpose of This Manual

Marine biology is a visual science. In order to appreciate that fully, you need to learn how to observe organisms from as many different perspectives as possible. The exercises in this manual are designed to develop your powers of accurate, critical observation and analysis to complement the lecture material in your marine biology course. Both laboratory and field exercises provide you with the opportunity to examine, in detail, a wide variety of marine (and estuarine) organisms in their natural habitat as well as in isolation. You will learn to identify marine organisms and relate their morphological adaptations to habitat and niche. You will also be exposed to a few of the quantitative methods that are used to describe species composition of marine communities.

Prelab and Pretrip Preparation

To make the most of the limited time available in the laboratory, or on field trips, please follow these guidelines:

1. *Prepare for lab.* Determine the topic to be covered (from the course schedule), then read the unit. Review the stated objectives of the exercise. Identify any concepts or procedural steps that you do not understand.

2. *Keep a journal.* Even though your instructor may not require one, it is helpful to make notes in a notebook (or "journal") for field trips and laboratory exercises. A journal can be designed for your personal use. It entitles you to record your ideas and observations without them being graded. When your ideas and observations have been recorded, they can be consulted at a later time and transcribed into a more formal style that may be required in Field and Laboratory Reports. The journal can also be used for your rough drafts of reports, and it can serve as a bibliography in which you record the detailed citation (author, date, title, publisher, page numbers, and so on) for each reference article or book that you use in the course.

3. *Use your resources.* Consult your text and any other references to find answers to the questions found in the manual, as well as those on the report and data sheets.

4. *Know what you'll need.* Bring essential items, such as a dissection kit, pencils, lab manual, and journal to the laboratory. Determine, from the procedural steps in the exercise, the need to bring any other items to class that day.

5. *Be careful.* Read the guidelines on safety at the end of this introduction.

Conduct in the Laboratory

To get good results from the exercises that you perform, please follow these guidelines:

1. *Cooperate with your peers.* The success of the team depends on all persons involved. It may be necessary to share and recycle reference books, specimens, and equipment. Return all items to their appropriate places when you are finished with them.

2. *Keep your equipment clean.* Success in cultivating and handling marine organisms depends on the cleanliness of the containers and utensils used. Many experiments have been ruined because of contamination of containers and dissecting instruments by the presence of unsuspected toxic substances.

a. **When experimenting with living forms, never use glassware that has been previously employed for some other purpose and which may have been in contact with toxic substances, such as formalin.**

b. Containers in a laboratory should be clearly labeled "live" or "preserved" to indicate their use for organisms. If not labeled, glassware should always be regarded with a certain degree of suspicion because it is impossible to know the purpose for which it was previously employed. All glassware, prior to its first use in the handling of living marine organisms, should be thoroughly washed in a biodegradable laboratory detergent, and rinsed no less than three times in distilled water. Subsequent cleaning may be done by rinsing in tap water followed by multiple rinsings in distilled water. Stack clean dishes upside down on clean toweling; direct contact of the open end of the dish with the laboratory table should be avoided.

c. Sea salts are corrosive. They destroy dissecting instruments as well as the metallic components of microscopes. Wipe up excess moisture as it occurs, and thoroughly clean and dry your instruments prior to storage.

3. *Follow the directions.* Follow the procedural sequence of the manual and do not skip around randomly through the manual while you work. Use the illustrations in the manual in the manner for which they were intended, that is, to complement the text; do not attempt to use them without consulting the text of the manual.

4. *Keep accurate records.* Since you may not have the opportunity to observe materials or habitats a second time, you need to observe critically and make complete and accurate records.

5. *Draw accurately and neatly.* Use the following checklist:

- The drawing is large enough to show features clearly.

- Only visible features are included.

- All parts are labeled (printed).

- All labels are aligned horizontally.

- Straight lines are used for labeling; lines do not cross over one another.

- Specimen name, scientific name, size, magnification, stain, condition, source, habitat, and so on,

are included.

6. *Check your work.* Before you leave the laboratory, verify that you accomplished the stated objectives, made all required observations, and completed data sheets.

Postlab and Posttrip Work

The following guidelines will help you organize your work after the lab period or field trip is over. As soon as possible after the laboratory session,

1. *Reread the unit.*

2. *Answer any required questions from the manual, including required interpretative questions.* Some of the questions can be answered only by deductive reasoning, and some may require outside reading. Use your notes, the manual, and other pertinent references. Some questions may be unanswerable and will stimulate independent study and/or discussion.

3. *Write the laboratory or field trip report.* Appendix 3 covers how to write reports.

SAFETY

In the **laboratory,** follow these basic rules of safety:

- Do not eat, drink, or smoke in the lab.

- Do not place personal articles on the floor or on laboratory benches.

- Learn the location and proper use of all safety equipment, such as fire extinguishers, fire blankets, eye-wash station, first-aid kits, and so on.

- Treat all chemical reagents with care. Follow the specific instructions of your instructor concerning their care and handling.

- Dispose of specimens, broken glassware, and chemicals in locations specified by the instructor.

- Notify the instructor immediately of any breakage, spill, or accident.

In the field, "safety first" is what you should remember. You may be working with unfamiliar equipment in unfamiliar situations and in often less than desirable weather conditions. Common sense must prevail in the field.

- KNOW what you are expected to do; if in doubt, ASK.

- Perform the task with full AWARENESS of those working with you, or near you.

- COMMUNICATE! Know who is carrying the first-aid kit. Report any injuries to your instructor.

Dangerous Organisms

Although few marine organisms fall into the "dangerous" category, you should be aware of those that do in the event you encounter such species. Venomous marine animals possess some form of apparatus that injects a venom into their victims. These structures range in size from the microscopic nematocysts of jellyfish to the spines of stingrays. Certain species of sponges, jellyfish, anemones, coral, gastropod and cephalopod mollusks, polychaete worms, sea urchins, bony fish and cartilaginous fish are venomous. *(Check with your instructor.)* Toxic organisms are those that contain or produce chemicals that can cause serious illness or death when they are ingested by predators, such as humans. Several species of fish are potentially toxic: barracuda, large jacks, groupers, puffer fish, and porcupine fish. When in doubt, ask.

While you are more apt to encounter nuisance species in tropical or subtropical waters, you should recognize the potential hazards of handling certain forms. Most of the hazards can be reduced or eliminated by wearing protective clothing (such as eyeware, gloves, shirts, pants, and sneakers). When necessary, you will be given specific warnings and instructions.

CONTENTS

Temperature, Salinity, Density, and Dissolved Oxygen

OBJECTIVES

After completing this unit, you will be able to

- Measure salinity and dissolved oxygen concentration of seawater;

- Understand the relationships between temperature, salinity, and density;

- Understand the relationships between temperature, salinity, and dissolved oxygen; and

- Interpret hydrographic data.

INTRODUCTION

Research in marine biology is concerned not only with the biological study of marine organisms but also with the organisms' ocean environment and their dependence on biotic and abiotic environmental factors. Thus, the science of marine biology includes ecological research that deals with descriptions and experimental analyses of biological processes in the ocean. Since a principal goal of this discipline is to describe the relationships between the organisms and their physical environment, it is not enough to concentrate solely on systematic surveys of the fauna and flora of the seas. It is also vitally important to describe quantitatively the marine environment of these organisms.

Any number of environmental parameters can be evaluated in the field and laboratory, often with extremely high precision. The choice of parameters, and the precision with which they are measured, depends upon the objectives of the study. Four parameters are measured routinely: temperature, salinity, density, and dissolved oxygen (DO) concentration.

Temperature

The activity, behavior, distribution, and survival of ectothermic marine organisms are controlled by the oceanic temperature range (-2 °C to 30 °C). Consequently, the metabolism (and many other activities) of these organisms varies with external temperature.

Vertical distribution of temperature, salinity, and density (Fig. 1.1) contribute to marine zonation. An obvious feature in most oceans is the **thermocline** (Fig. 1.1a), a zone located beneath the surface in which a rapid decrease in temperature occurs relative to the decrease in depth. The decrease in temperature is usually accompanied by dramatic increases in salinity **(halocline)** (Fig. 1.1b) and density **(pycnocline)** (Fig. 1.1c). The large density difference on either side of the thermocline (and pycnocline) effectively separates the oceans into two layers. This "barrier" impedes the exchange of gases, nutrients, and organisms between the two layers.

Salinity

Marine organisms live in an environment that includes a vast array of dissolved salts. The concentration of the salts is expressed as **salinity.** The formal definition of salinity is complex and somewhat intimidating:

> . . . the weight of solid materials in grams *(in vacuo)* contained in one kilogram of water when all the carbonate has been oxidized, all the bromide and iodide have been replaced by chloride, all organic matter has been oxidized, and the residue has been dried at 480 °C to constant weight.

Fortunately, the practicing marine biologist can set aside that cumbersome definition and work with more easily understood conventions that relate salinity to the chemistry of seawater. Salinity refers to the amount of dissolved solids in seawater, and is stated in units of **parts per thousand,** abbreviated as **ppt** or as **o/oo.** While the latter notation might seem foreign to you, recall that parts per hundred equals percent and is abbreviated o/o.

1

The concept of salinity results from scientists' attempts to identify a simple analytical procedure that reliably represents the chemical constitution of any given sample of seawater. The ratios of the **major dissolved ions** (Na^+, Mg^{++}, Ca^{++}, K^+, Cl^-, SO_4^{--}, HCO_3^-) in seawater are constant. Thus, it is possible to calculate the concentrations of all of these ions from the known concentration of any one of them.

Of the major ions, the concentration of Cl^- is the easiest to analyze with reasonable accuracy. The chloride concentration is known as **chlorinity,** and has the following relationship with salinity:

Salinity = 1.80655 chlorinity (o/oo)

Open ocean salinity is approximately 35 o/oo, which means that there are 35 grams of dissolved solids in each 1000 grams of water. Near shore, however, seawater is diluted by freshwater from river discharge. In the mouth of Chesapeake Bay, for example, the salinity is 30 o/oo. In the upper end of that estuary, the salinity is less than 1 o/oo.

Marine organisms must maintain reasonably constant internal environmental conditions with respect to salt and water balance. Some species **conform** to the environment; their internal ionic composition is similar to, and varies with, changes in external salinity. Other organisms **regulate** their internal environments by selectively absorbing some ions and preventing others from entry. The distribution of species in the marine environment depends upon the ability of each species to tolerate changing conditions of salinity.

Density

Density is a measurement of the specific gravity of water, and is usually expressed as grams per milliliter (g/mL); however, by convention, oceanographers usually omit these units when discussing density measurements. For example, density of pure water is 1.000; that of open ocean water (salinity = 35 o/oo) is 1.025. Oceanographers have further modified their use of density values by converting them to **sigma-t** (σ_t) values:

Sigma-t = (density − 1)1000

This transformation makes numerals beyond the third decimal place more manageable. For example, the density 1.02511 becomes 25.11 upon conversion to its sigma-t value (Fig. 1.2).

Density increases curvilinearly with a temperature decrease and linearly with a salinity increase (Fig. 1.3). Observe that density increases as the water becomes colder and more saline.

Dissolved Oxygen

Oxygen dissolved in water is as essential to aquatic organisms as gaseous atmospheric oxygen is to terrestrial forms. While the atmosphere is an important source of dissolved oxygen (DO), phytoplankton, macroscopic algae and higher marine plants also increase the amount of DO in the water.

Oxygen is generally not a limiting factor in the ocean; seawater normally contains between 5 and 14 parts per million (ppm) (= mg/L) DO. The survival of many organisms is threatened on those few

Figure 1.1 Typical temperature, salinity, and density profiles of the ocean. Decreasing temperature and increasing salinity combine to intensify the pycnocline.

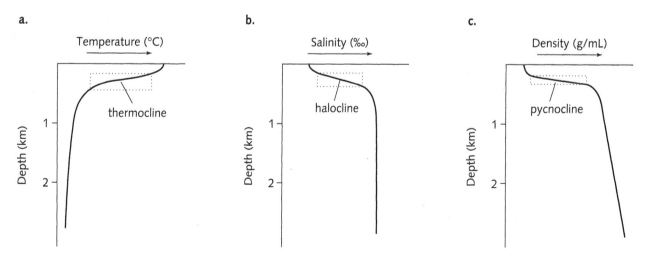

occasions when the concentration decreases below that range (conditions become **hypoxic).** Oxygen content can be depleted in a variety of ways. For example, increased temperature increases the metabolism of most aquatic animals, causing them to consume more oxygen per unit of time than at a lower temperature. Higher temperature also de-creases the solubility of oxygen in water, and thus reduces its availability to aquatic organisms. Bacterial decomposition consumes enormous quantities of oxygen, and can cause the system to rapidly become **anoxic** (no oxygen) if it occurs in warm water.

The amount of oxygen that can dissolve in water depends primarily on temperature and salin-

Figure 1.2 Relation between salinity, chlorinity, and density at five temperatures. Sigma-t = (density − 1) × 1000.

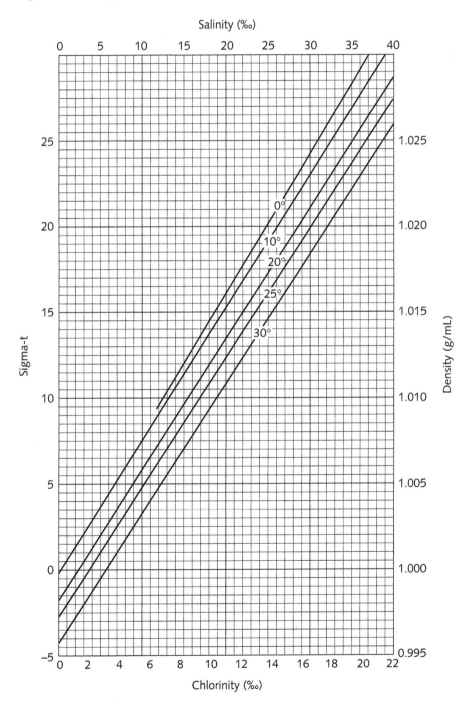

ity (see Fig. 1.4 and Table 1.1). Because these relationships are well defined, oxygen solubility can be predicted from specially designed plots, called **nomographs,** which we will use later in the lab (see Fig. 1.8). The calculated values are theoretical—that is, they represent **expected** conditions based on a given set of parameters. It is important to recognize that two bodies of water with the same salinity and temperature may not contain the same **observed** DO concentration. If the DO concentration equals the expected solubility, the water is said to be fully saturated **(100% saturation).** Biological respiration can reduce DO concentration to values below saturation level. On the other hand, excessive photosynthetic activity can produce **supersaturated** conditions (greater than 100%).

In this unit you will have the opportunity to measure the temperature, salinity, density, and dissolved oxygen of water samples. You will also work with typical or simulated **hydrographic data** (measurements of environmental parameters at sampling stations).

Figure 1.3 Variation in density with temperature (a) and salinity (b).

a.

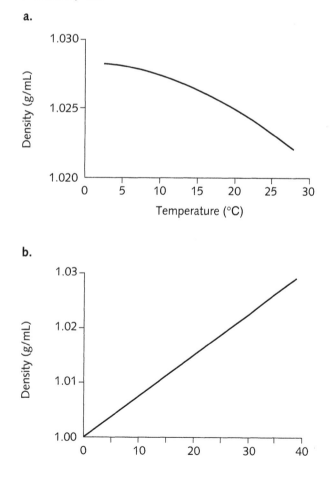

b.

Figure 1.4 Effect of water temperature and salinity on dissolved oxygen concentration (DO). In (a), the curvilinear relationship is illustrated for three representative salinities. In (b), the linear relationships are shown for four temperatures.

a.

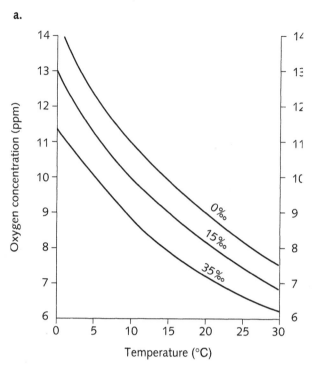

b.

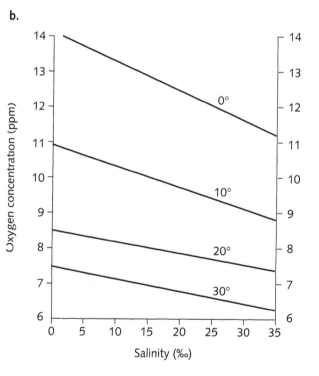

Table 1.1 Oxygen solubility (mg/L) in seawater from a water-saturated atmosphere. Solubility is based on temperature (°C) and salinity (o/oo) values. Oxygen solubility in ppm (mL/L) can be determined by multiplying mg/L by 0.7.

Temp. (°C)	Salinity (ppt)								
	0	5	10	15	20	25	30	35	40
5	14.8	14.4	13.9	13.5	13.0	12.5	12.1	11.6	11.2
10	13.0	12.6	12.2	11.8	11.4	11.0	10.6	10.2	9.8
15	10.3	10.0	9.7	9.4	9.2	8.9	8.6	8.3	8.1
20	9.4	9.1	8.8	8.6	8.4	8.1	7.9	7.6	7.4
25	8.5	8.3	8.0	7.8	7.6	7.4	7.2	6.9	6.7
30	7.8	7.6	7.4	7.2	7.0	6.8	6.6	6.4	6.2

I. ANALYTICAL PROCEDURES

A. Determination of Salinity

EXERCISE 1

Chemical Analysis of Seawater

Until recently, the Mohr titration was a standard chemical method for determining chloride concentration. It was based on the principles that (1) silver and chloride ions react to form an insoluble white precipitate (AgCl), and (2) silver and chromate ions form a somewhat more soluble, red-orange precipitate (Ag_2CrO_4). When a solution of silver ions is continuously added to a solution containing both chloride and chromate ions, AgCl will precipitate until all the chloride is removed from the solution. At that instant, silver chromate will begin to precipitate.

Knudsen applied these principles to the development of an accurate technique for the measurement of chloride ion in seawater. This technique, the Knudsen–Mohr titration, which has a precision of ± 0.02 o/oo salinity, was the standard method used by oceanographers from 1900 until the early 1960s. Because it was difficult to perform in the field, the Knudsen–Mohr titration was supplanted by more sophisticated, but simpler to use methods (discussed in Exercise 2).

You may be asked to perform a chemical analysis of your samples. If so, you will be provided with instructions on how to carry out this analysis. Details of the titration procedure may be found in Strickland and Parsons (1972) or *Standard Methods* (1989).

EXERCISE 2

Determining the Electrical Conductivity of Seawater

The Knudsen–Mohr titration has been replaced by the **induction salinometer** (conductivity meter), which measures the electrical conductivity of seawater (with a precision of ± 0.003 o/oo). This method's reliability in determining salinity is based upon the known relationship between electrical conductivity and salinity (Fig. 1.5). Because both values vary with temperature, these parameters must be measured simultaneously. The salinometer takes both measurements and compensates for temperature. This instrument measures the resistance (to an electric current) of the seawater sample in a Wheatstone Bridge circuit within the instrument. The device is calibrated to provide a direct read-out of temperature and either salinity or conductivity.

PROCEDURE

1. Become familiar with the controls on the salinometer you will use. There will either be a direct read-out display, or three scales on the instrument: temperature (C), salinity (o/oo) and conductivity (μmhos = micromhos).

   Because several models of salinometers are available, you will need to consult your instrument's instruction manual for operating details.

2. The probe (sensing unit) connects to the meter by an electrical cable.

CAUTION

The probe is sensitive and should not be handled roughly or dropped.

3. After you have tested and calibrated the salinometer, place the probe in the water sample. If you are sampling in the field, lower the probe to the desired depth. The water depth may be easily observed if the cable's length is marked in meters. Gentle agitation of the probe in the sam-

ple keeps water flowing over the electrodes and ensures good results.

4. Read the temperature.

5. If necessary, adjust the instrument for the temperature you just measured.

6. Read the salinity (or conductivity) and record your data on the appropriate Station Log or in your lab notebook.

7. Repeat steps 3 through 6 for each sample to be analyzed.

8. When you finish with the instrument, rinse the probe in freshwater and blot it dry. Store it in a plastic bag. Disconnect the cable from the meter (if possible), and coil it for storage.

EXERCISE 3

Calculating Salinity from Density of Seawater

Salinity and temperature affect the density of water in a highly predictable fashion (see Figs. 1.2 and 1.3). Thus, salinity can be calculated from density of seawater at a known temperature. **Hydrometers** (calibrated glass floats) are used to measure the density. The higher the hydrometer floats (buoyed), the denser the water. A conversion table allows you to calculate salinity from the observed density and temperature readings. This procedure is simple, quick, and inexpensive, but less accurate than other methods.

PROCEDURE

1. Use the Density Worksheet found on page 9.

2. Place the seawater sample in a 500-mL graduated cylinder, or a hydrometer jar (400 mm high with a 45-mm inside diameter). Fill the container two-thirds full of the sample.

3. Measure the temperature of the sample. Record this value in column B on the worksheet.

4. Place the hydrometer in the sample, and measure the density by reading the value on the hydrometer scale that corresponds with the meniscus of the water. Record the reading in column C.

5. Correct the observed density value to the corresponding value at 15 °C by referring to Table 1.2

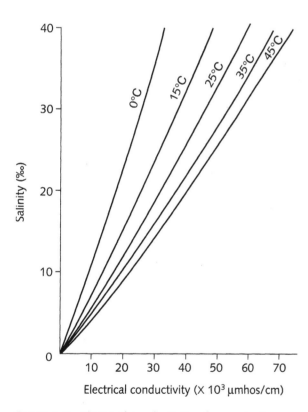

Figure 1.5 Electrical conductivity of seawater is a function of temperature and salinity. Conductivity is expressed in units of micromhos (μmhos), indicating that it is the reciprocal of resistance, measured as micro-ohms.

Table 1.2 Temperature Correction Table (Density Measurements)

Measured Density	Measured Water Temperature (°C)																				
	0	1	2	3	4	5	6–10	11	12	13–17	18	19–21	22	23	24	25	26	27	28	29	30
1.000	1.000	1.000	.999	.999	.999	.999	.999	1.000	1.000	1.000	1.000	1.001	1.001	1.001	1.002	1.002	1.002	1.002	1.003	1.003	1.003
1.001	1.001	1.000	1.000	1.000	1.000	1.000	1.000	1.001	1.001	1.001	1.001	1.002	1.002	1.002	1.003	1.003	1.003	1.003	1.004	1.004	1.004
1.002	1.002	1.001	1.001	1.001	1.001	1.001	1.001	1.002	1.002	1.002	1.003	1.003	1.003	1.003	1.004	1.004	1.004	1.004	1.005	1.005	1.005
1.003	1.002	1.002	1.002	1.002	1.002	1.002	1.002	1.003	1.003	1.003	1.004	1.004	1.004	1.005	1.005	1.005	1.005	1.005	1.006	1.006	1.006
1.004	1.003	1.003	1.003	1.003	1.003	1.003	1.003	1.004	1.004	1.004	1.005	1.005	1.005	1.006	1.006	1.006	1.006	1.006	1.007	1.007	1.007
1.005	1.004	1.004	1.004	1.004	1.004	1.004	1.004	1.004	1.005	1.005	1.006	1.006	1.006	1.007	1.007	1.007	1.007	1.007	1.008	1.008	1.008
1.006	1.005	1.005	1.005	1.005	1.005	1.005	1.005	1.005	1.006	1.006	1.007	1.007	1.007	1.008	1.008	1.008	1.008	1.008	1.009	1.009	1.009
1.007	1.006	1.006	1.006	1.006	1.006	1.006	1.006	1.006	1.007	1.007	1.008	1.008	1.008	1.009	1.009	1.009	1.009	1.009	1.010	1.010	1.010
1.008	1.007	1.007	1.007	1.007	1.007	1.007	1.007	1.007	1.008	1.008	1.009	1.009	1.009	1.010	1.010	1.010	1.010	1.010	1.011	1.011	1.011
1.009	1.008	1.008	1.008	1.008	1.008	1.008	1.008	1.008	1.009	1.009	1.010	1.010	1.010	1.011	1.011	1.011	1.011	1.011	1.012	1.012	1.012
1.010	1.009	1.009	1.009	1.009	1.009	1.009	1.009	1.009	1.010	1.010	1.011	1.011	1.011	1.012	1.012	1.012	1.012	1.012	1.013	1.013	1.013
1.011	1.010	1.010	1.010	1.010	1.010	1.010	1.010	1.010	1.011	1.011	1.012	1.012	1.012	1.013	1.013	1.013	1.013	1.013	1.014	1.014	1.014
1.012	1.011	1.011	1.011	1.011	1.011	1.011	1.011	1.011	1.012	1.012	1.013	1.013	1.013	1.014	1.014	1.014	1.014	1.014	1.015	1.015	1.015
1.013	1.012	1.012	1.012	1.012	1.012	1.012	1.012	1.012	1.013	1.013	1.014	1.014	1.014	1.015	1.015	1.015	1.015	1.015	1.016	1.016	1.016
1.014	1.013	1.013	1.013	1.013	1.013	1.013	1.013	1.013	1.014	1.014	1.015	1.015	1.015	1.016	1.016	1.016	1.016	1.016	1.017	1.017	1.017
1.015	1.014	1.014	1.014	1.014	1.014	1.014	1.014	1.014	1.015	1.015	1.016	1.016	1.016	1.017	1.017	1.017	1.017	1.017	1.018	1.018	1.018
1.016	1.015	1.015	1.015	1.015	1.015	1.015	1.015	1.015	1.016	1.016	1.017	1.017	1.018	1.018	1.018	1.019	1.019	1.019	1.019	1.020	1.020
1.017	1.016	1.016	1.016	1.016	1.016	1.016	1.016	1.016	1.016	1.017	1.018	1.018	1.019	1.019	1.019	1.020	1.020	1.020	1.020	1.021	1.021
1.018	1.017	1.017	1.017	1.017	1.017	1.017	1.017	1.017	1.017	1.018	1.019	1.019	1.020	1.020	1.020	1.021	1.021	1.021	1.021	1.022	1.022
1.019	1.018	1.018	1.018	1.018	1.018	1.018	1.018	1.018	1.018	1.019	1.020	1.020	1.021	1.021	1.021	1.022	1.022	1.022	1.022	1.023	1.023
1.020	1.019	1.019	1.019	1.019	1.019	1.019	1.019	1.019	1.019	1.020	1.021	1.021	1.022	1.022	1.022	1.023	1.023	1.023	1.023	1.024	1.024
1.021	1.020	1.020	1.020	1.020	1.020	1.020	1.020	1.020	1.020	1.021	1.022	1.022	1.023	1.023	1.023	1.024	1.024	1.024	1.024	1.025	1.025
1.022	1.021	1.021	1.021	1.021	1.021	1.021	1.021	1.021	1.021	1.022	1.023	1.023	1.024	1.024	1.024	1.025	1.025	1.025	1.025	1.026	1.026
1.023	1.021	1.021	1.022	1.022	1.022	1.022	1.022	1.022	1.022	1.023	1.024	1.024	1.025	1.025	1.025	1.026	1.026	1.026	1.026	1.027	1.027
1.024	1.022	1.022	1.022	1.023	1.023	1.023	1.023	1.023	1.023	1.024	1.025	1.025	1.026	1.026	1.026	1.027	1.027	1.027	1.027	1.028	1.028
1.025	1.023	1.023	1.023	1.023	1.024	1.024	1.024	1.024	1.024	1.025	1.026	1.026	1.027	1.027	1.027	1.028	1.028	1.028	1.028	1.029	1.029
1.026	1.024	1.024	1.024	1.024	1.024	1.025	1.025	1.025	1.025	1.026	1.027	1.027	1.028	1.028	1.028	1.029	1.029	1.029	1.029	1.030	1.030
1.027	1.025	1.025	1.025	1.025	1.025	1.026	1.026	1.026	1.026	1.027	1.028	1.028	1.029	1.029	1.029	1.030	1.030	1.030	1.030	1.031	1.031
1.028	1.026	1.026	1.026	1.026	1.026	1.027	1.027	1.027	1.027	1.028	1.029	1.029	1.030	1.030	1.030	1.031	1.031	1.031	1.031	1.032	1.032
1.029	1.027	1.027	1.027	1.027	1.027	1.027	1.028	1.028	1.028	1.029	1.030	1.030	1.031	1.031	1.031	1.032	1.032	1.032	1.032	1.033	1.033
1.030	1.028	1.028	1.028	1.028	1.028	1.028	1.029	1.029	1.029	1.030	1.031	1.031	1.032	1.032	1.032	1.033	1.033	1.033	1.033	1.034	1.034
1.031	1.029	1.029	1.029	1.029	1.029	1.029	1.030	1.030	1.030	1.031	1.032	1.032	1.033	1.033	1.033	1.034	1.034	1.034	1.034	1.035	1.035

Density of Water at 15 °C

Note: This table provides temperature-corrected densities at 15 °C of water densities measured at other temperatures. For example, if a water sample having a measured density of 1.023 at 22 °C were cooled to 15 °C, the density would be 1.025.

For more accurate results adjust the water temperature of your sample to 15 °C before making a hydrometer reading rather than use this table.

with the two measured values. Record the corrected (to 15 °C) density in column D.

6. Determine the salinity of your sample by referring to Table 1.3 with the corrected density value. Record this value in column E and transfer your final salinity values to your Station Log or your lab notebook.
   Example: Temperature of sample (measured in lab) = 21.2 °C. Observed (measured) density of sample = 1.019. Refer to Table 1.2 at the measured reading of 1.019. Find the value of 1.020 in the temperature column labeled 19–21°. Refer to Table 1.3 with the density reading of 1.020, and read salinity of the sample as 27 o/oo.

EXERCISE 4

Determining the Refractive Index of Seawater

Refractometry is an accurate, simple, and rapid method for measuring salinity. It is based on the principle of light refraction, or bending of light waves, as light passes from an optically thinner medium (such as air) to an optically thicker medium (seawater). The degree of refraction (recorded as the **refractive index**) depends on the wavelength of light and increases with increasing salinity and decreasing temperature (see Fig. 1.6 and Table 1.4).

The refractive index (ranging from less than 1.3325 to greater than 1.3425) can be measured with an optical device called a **refractometer** (Fig. 1.7). Basic models provide a read-out as the refractive index only, and do not permit temperature compensation. Some models offer the additional options of reading specific gravity (from 1.000 to 1.070, in 0.001 increments) or salinity (with accuracy to 1 o/oo), and provide automatic temperature compensation between 10 °C and 30 °C.

PROCEDURE

1. *Calibrate* the refractometer.

   a. Place a few drops of distilled water on the face of the prism (Fig. 1.7), and close the cover plate.

   b. Aim the prism end of the instrument at a bright light.

   c. Rotate the eyepiece while looking through it until the image focuses and the scale becomes visible.

Table 1.3   Density–Salinity Conversion Table 15 °C (59 °F)

Density (at 15 °C)	Salinity (o/oo)	Density (at 15 °C)	Salinity (o/oo)
0.999	0	1.016	22
1.000	1	1.017	23
1.001	2	1.018	25
1.002	4	1.019	26
1.003	5	1.020	27
1.004	6	1.021	29
1.005	8	1.022	30
1.006	9	1.023	31
1.007	10	1.024	32
1.008	12	1.025	34
1.009	13	1.026	35
1.010	14	1.027	36
1.011	15	1.028	38
1.012	17	1.029	39
1.013	18	1.030	40
1.014	19	1.031	42
1.015	21	1.032	43

Note: Use Table 1.2 to obtain temperature-corrected densities before using this table. Other factors beside temperature and salinity can affect water density. Salinity values obtained with this table should be considered only approximations.

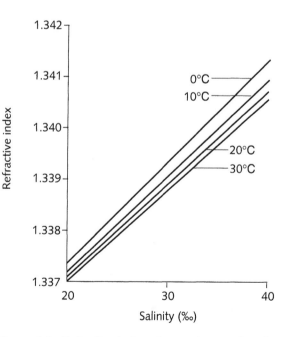

Figure 1.6   Refractive index of seawater as a function of salinity and temperature.

d. Read the scale where the sharp boundary line (between blue and white, or light and dark fields) crosses the scale values. For distilled water, this value should be 1.000 on the specific gravity scale, or 1.333 on the refractive index scale. Adjust the line to the correct value by turning the scale adjustment screw.

e. Open the cover plate and dry the prism surface.

2. Determine the salinity of your sample.

a. Place a drop of your unknown sample on the prism surface and close the cover plate.

b. Focus the image and read the value indicated by the boundary line. Make the appropriate compensation for temperature. The refractometer may make this adjustment automatically, or you may correct your reading with a temperature compensation chart (consult the operating manual for the instrument). Record your reading. If you cannot read salinity directly, find the measured refractive index in Table 1.4. By extrapolation, you will be able to estimate the salinity of your sample. If your instrument provides you with the specific gravity (= density), use Table 1.3 to determine salinity from density.

Density Worksheet Determining Salinity from Density (See procedure on page 6)

A	B	C	D	E
Sample number	Temp. (°C)	Hydrometer reading	Density corrected to 15 °C (from Table 1.2)	Salinity (o/oo) (from Table 1.3)

Table 1.4 Refractive Index of Seawater

S (o/oo)	Temperature (°C)					
	0	5	10	15	20	25
0	1.3 3395	1.3 3385	1.3 3370	1.3 3340	1.3 3300	1.3 3250
5	3500	3485	3465	3435	3395	3345
10	3600	3585	3565	3530	3485	3435
15	3700	3685	3660	3625	3580	3525
20	3795	3780	3750	3715	3670	3620
25	3895	3875	3845	3805	3760	3710
30	3991	3966	3935	3898	3851	3798
31	4011	3985	3954	3916	3869	3816
32	4030	4004	3973	3934	3886	3834
33	4049	4023	3992	3953	3904	3851
34	4068	4042	4011	3971	3922	3868
35	4088	4061	4030	3990	3940	3886
36	4107	4080	4049	4008	3958	3904
37	4127	4099	4068	4026	3976	3922
38	4146	4118	4086	4044	3994	3940
39	4166	4139	4105	4062	4012	3958
40	(4185)	(4157)	(4124)	(4080)	(4031)	(3976)
41	(4204)	(4176)	(4143)	(4098)	(4049)	(3944)

Figure 1.7 Refractometer.

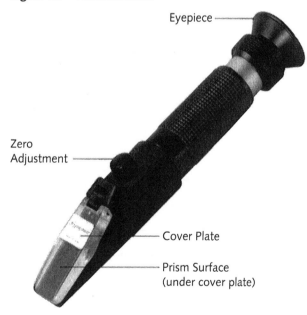

Eyepiece

Zero
Adjustment

Cover Plate

Prism Surface
(under cover plate)

B. Determination of Dissolved Oxygen

E X E R C I S E   1

Collection of Water Samples

Proper sampling is critical to oxygen determination. If you do not measure DO directly with a polarigraphic oxygen electrode, you must collect the sample in a way that prevents contact of the water with the atmosphere. If you plan to titrate the samples, you must treat them chemically (fix them) prior to titration.

PROCEDURE

1. The sample bottle must have a conical stopper or lid that will not trap air bubbles when it is inserted in the full bottle. Biochemical oxygen demand (BOD) bottles are commonly used.

2. Attach a piece of glass tube slightly longer than the height of the bottle, fire-polished at both ends, to approximately 20 cm of rubber tubing. Attach the free end of the rubber tubing to the outlet faucet of the sampling device.

3. Hold the glass tube in a vertical position, and open the faucet or pinch clamp to allow the water to displace the air in the tubing.

4. Insert the glass tube into the sample bottle so that the end touches the bottom. Allow the water to flow into the sample bottle until about 50 mL of the sample overflows. This action expels the part of the sample that was in contact with air while the bottle was filling. With the water still running, slowly remove the glass tube from the bottle and quickly insert the cap *without trapping any air bubbles.*

5. Turn the bottle upside down. If you see air bubbles, discard the sample and fill the bottle again.

6. Label the bottle or record the number of the bottle in your collection log with all other pertinent information about the sample collection (such as date, location, depth, and temperature).

7. Analyze the sample immediately with the DO meter, or "fix" it chemically if you intend to do a Winkler titration (see Exercise 2) .

EXERCISE 2
Chemical Analysis

The Winkler titration (Strickland and Parsons, 1972; *Standard Methods*, 1989) is based on the formation of a precipitate of manganous hydroxide. During chemical fixation, potassium iodide (KI), sodium hydroxide (NaOH), and manganous sulfate ($MnSO_4$) are added to the sample. The oxygen dissolved in the water is rapidly absorbed by the manganous hydroxide that precipitates, and forms a flocculent (a suspension of loosely aggregated material). The flocculent acts as a gathering agent for oxygen. Upon acidification in the presence of iodide during the fixing procedure, iodine is released in a quantity equivalent to the DO present. The liberated iodine is then titrated with a standard sodium thiosulfate solution, using starch as the indicator.

EXERCISE 3
Using a Dissolved Oxygen Meter

Measuring DO content of water is more convenient with a DO meter, an electronic device that incorporates a specialized sensing probe. The probe is a cylinder with a thin, permeable membrane (Teflon or polyethylene) covering polarigraphic sensors, and thermistors that provide temperature measurement and compensation. The membrane isolates the sensor elements from the environment, but allows oxygen to enter. When a polarizing voltage is applied across the sensor, oxygen that has passed through the membrane reacts at the cathode, causing a current to flow. The membrane passes oxygen at a rate proportional to the pressure difference on either side of the membrane. Thus, the amount of oxygen that diffuses across the membrane is proportional to the amount of oxygen outside the membrane.

Probes are available in a variety of shapes and sizes. Some fit in the neck of a standard BOD (biochemical oxygen demand) water sample bottle, and contain automatic stirrers. Others are larger and may be used to read DO directly in the water at any selected depth. They may or may not have temperature sensors.

The procedure for the instrument you will use may differ somewhat from the brief description that follows. The probe may be ready to use, and the meter may be calibrated automatically. If you need to perform these procedures, consult the operating manual for detailed instructions.

PROCEDURE

1. Examine a DO meter. The meter may have two scales, one for DO (reading in ppm, or mg/L) and one for temperature (°C). Function knobs allow you to calibrate the instrument and make the appropriate compensation for salinity.

2. Locate the probe, *but do not touch* the membrane covering the gold cathode and the chamber (containing KCl) of the sensor probe. Make certain that no air bubbles are trapped under the membrane. (If you need to prepare the probe for use and/or perform the calibration, consult the instrument's operating manual.)

3. Place the probe in the sample and then stir the sample.
   NOTE: Water must flow past the probe at a rate of at least 1 ft/s. If the probe does not have a built-in stirrer, put a stirring bar in the bottom of the sample bottle and use a magnetic stirrer.

4. Read the temperature of the water sample, and record the value on your data sheet.

5. If necessary, compensate the instrument for the temperature and salinity of the water sample.

6. Read the dissolved oxygen of the sample. Keep the probe agitated (see Step 3). Read and record the DO value.

7. Repeat Steps 3–6 for each water sample.

8. When you finish taking readings, insert the probe into the calibration chamber, but *do not* turn off the instrument until you are instructed to do so.

E X E R C I S E 4

Determination of Oxygen Solubility

PROCEDURE

1. Measure the temperature, salinity, and DO of your water samples. Use the methods described earlier in this exercise.

2. Determine oxygen solubility for your samples from the **nomograph** in Figure 1.8.

   a. Position a ruler (at least 12 inches long) on the nomograph so that the measured values of salinity and temperature are in line. The ruler will extend across the oxygen scale. Read the oxygen solubility (in mg/L) at that point, which is the oxygen solubility that can be *expected* to occur at one atmosphere of pressure at the measured salinity and temperature.
   NOTE: Oxygen concentration can be expressed as ppm (mg/L) or as ppt (mL/L). Use these equations to convert one to the other:

   $$(mg/L) \times 0.7 = mL/L$$

   $$(mL/L) \times 1.428 = mg/L$$

   b. Determine the percentage saturation of oxygen of your sample with the equation:

   % saturation = (measured DO)/(expected DO)(100)

   Example Determinations: For sample P, the ruler is aligned with 33 o/oo and 15 °C. The expected DO value of 8 ppm appears where the ruler intersects the DO line. Dividing 6ppm (measured value) by 8 ppm gives 0.75. Thus, 6 ppm is 75% of the 8 ppm value.

Sample	Measured Values			Expected DO (ppm)	Saturation (%)
Sample	DO (ppm)	Salinity (o/oo)	Temperature (°C)		
P	6	33	15	8.0	75
Q	6	20	18	8.2	73
R	6	15	11	9.8	61

Figure 1.8 Nomograph of oxygen solubility in seawater. A straight line constructed between any two variables will intersect the third scale at the appropriate value of the third variable. Oxygen solubility in ppt (mL/L) can be obtained by multiplying mg/L by 0.7.

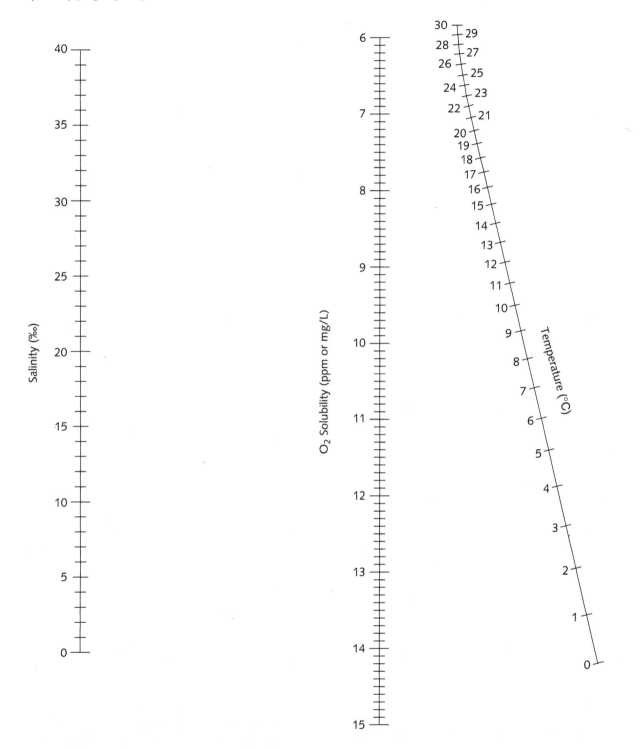

II. APPLICATIONS OF TEMPERATURE, SALINITY, AND DISSOLVED OXYGEN DATA

EXERCISE 1

Comparison of Methods in Determination of Salinity

PROCEDURE

Make several dilutions of a stock solution of sea-water and measure the salinity of each by the hydrometer, refractometer, salinometer, and titration methods.

Interpretation

1. Construct a table that compares your results. Be certain to include a descriptive title. Consult Appendix II for guidelines to table construction.

2. Identify discrepancies and attempt to explain unexpected variations or deviations from the expected results.

3. Describe the advantages and disadvantages of each method.

EXERCISE 2

Effects of Temperature and Salinity on Dissolved Oxygen

PROCEDURE

1. Obtain a set of nine prepared water samples (unknowns): three different salinities, aerated, and stored at three different temperatures.

2. Using the relevant procedures from the "Analytical Procedures" section of this unit, measure and record the salinities and DO concentrations of the unknown samples.

Interpretation

1. Plot DO concentration in relation to temperature.

2. Describe the relation of DO to temperature and salinity.

3. Plot DO concentration in relation to salinity.

4. Determine the expected oxygen solubility from the temperature and salinity measurements.

5. Do your observations agree with the expected values? Can you explain any discrepancies between them?

EXERCISE 3

Simulated Field Data

PROCEDURE

1. Obtain a series of water samples for analysis. They may be prepared by the instructor, but you should assume that they were collected from designated depths at one or more offshore sampling stations.

2. Record, on the Hydrographic Station Log (page 17), all pertinent data accompanying the samples (for example, location, date and time of sampling, and the temperature of the water at each depth at the time of sampling).

3. Using the relevant procedures from the "Analytical Procedures" section of this unit, determine the **salinity** and **dissolved oxygen** of the samples. Make the necessary calculations and record your final determined values on the Station Log.

4. Calculate the **density** of each sample from (a) the value of your salinity determination, and (b) the temperature provided for you (sample data included):

   a. Find your salinity value (e.g., 31 o/oo) in Table 1.3, and record the corresponding density (e.g., 1.023) at 15 °C.

   b. In Table 1.2, locate a density reading of 1.023 that corresponds to the temperature of your simulated field sample (e.g., 22 °C). Record the density from the first column in the table (in this case, 1.025) in your Station Log. For example, a sample of water at 22 °C with a density of 1.021 would have a density of 1.023 at 15 °C.

5. Determine the **percentage saturation** of the observed DO values through the methods described for determining oxygen solubility in seawater (page 12). Record these data on the Station Log.

6. Graph the station data. (See Appendix II for an example.)

   a. For each station, plot **temperature** in relation to **depth.** Construct the graph so that depth lies on the left axis, with the surface at the top and the bottom depth at the bottom. Temperature values should be placed on the horizontal axis at the top of the page, with values increasing from left to right.

   b. Plot **salinity** in relation to **depth** in a similar fashion. Both temperature and salinity may be included in the same plot if the axes are carefully arranged and clearly labeled.

   c. Plot **density** in relation to **depth.** It may be possible to place the density plot on the same graph with salinity and temperature, but ask your instructor first.

   d. On a separate graph, plot **DO** concentration in relation to **depth** as described for salinity and temperature.

   e. Plot **percentage oxygen saturation** in relation to **depth.** These data may be plotted on the same graph as DO if the graph is carefully planned and the axes are clearly labeled.

Interpretation

Write a narrative report in which you analyze your data and draw conclusions from them. Refer to your textbook for assistance. Consider the following suggestions and questions as you deliberate your findings.

1. Examine your plots and indicate if the results (trends and patterns) are what you would expect, based on the station information provided.

2. With respect to temperature and salinity relationships, can you detect the presence of a thermocline or a halocline? If they are present, do they reinforce each other by creating a well-defined pycnocline?

3. Describe the observed patterns of DO concentration and oxygen saturation relative to depth. Are they what you would expect, given the temperature and salinity data? Where is the saturation highest? Why? Do you detect the presence of an oxygen minimum layer?

REFERENCES

Strickland, J. D. H. and T. R. Parsons. 1972. *A practical handbook of seawater analysis.* Bull. 167. 2nd. ed. Fisheries Res. Bd. Canada, Ottawa, Ontario.

Standard Methods for the Examination of Water and Wastewater. 1989. 17th ed. American Public Health Assoc., Washington, D.C. 1644 pp.

NAME _____ SECTION _____ DATE _____

HYDROGRAPHIC STATION LOG

Station No. _____ Location: Lat _____ Long
_____

Date _____ Time on Station _____ Air Temp (C)
_____

Wind: Dir _____ Vel _____ Sea State _____

Depth (M)	Temp (C)	Sample No.	Density	Salinity (o/oo)	Oxygen (ppm) Obs Exp	% Sat
		1				
		2				
		3				
		4				
		5				
		6				
		7				
		8				
		9				
		10				
		11				
		12				
		13				
		14				
		15				

Comments:

Marine Invertebrate Classification and Identification

OBJECTIVES

After completing this unit, you will be able to

- Understand the basic principles of taxonomic classification and phylogeny;

- Use taxonomic keys and guides to identify marine invertebrates; and

- Recognize the major characteristics of the phylum and class of select groups of marine invertebrates.

INTRODUCTION

The marine environment is inhabited by a bewildering array of organisms. In fact, there are more than one million different kinds (species) of animals, 95 percent of which are invertebrates classified into 33 phyla (Fig. 2.1). Most of the invertebrate phyla are predominantly or exclusively marine.

Organisms are identified and named through the process of biological classification. First, the morphology and other features (characters) of the organisms are described. Then, the organisms are ordered and ranked into groups by analysis of patterns in the described characters. Specimens that share a large number of characters are grouped into **species** (the word is both singular and plural). Related species are placed into groups called **genera** (singular, genus), similar genera into **families,** and so on, up to the higher categories of subkingdom and kingdom. Such categories are hierarchic (for example, phyla include subphyla, which in turn include classes). Groups above species level are referred to as **taxa** (singular, taxon); these taxa are the subject matter of **taxonomy**—the systematic process that Carolus Linnaeus originated in 1735. Taxonomy involves three major activities: **classification, nomenclature,** and **identification.**

Biological Classification

In biological classification, species are studied to find common, homologous (similar structures attributable to common origin) features that can be used to group species into taxa. Careful observation reveals basic patterns of form and function within the observed diversity. This evaluation of shared characteristics provides insight into evolutionary relationships among taxa, which biologists then use to classify organisms into schemes that reflect those relationships, or **phylogeny** (Fig. 2.1). As new evidence accumulates, these schemes may change. Not only can systematists reassign species to different taxa, but they can also change the taxa names. Students of biology must be aware of the dynamic nature of taxonomy, and learn to cope with these variants. As a base point of taxonomic reference, the invertebrate classification scheme used in this manual follows that used by Brusca and Brusca (1990).

Nomenclature

The naming of animal species is controlled by a complex set of rules: *The International Code of Zoological Nomenclature.* The code is derived from Carolus Linnaeus, who initiated the **binominal system** by which every species has two parts to its scientific, or species, name. For example, examine the Linnaean hierarchy for the edible blue crab of the Chesapeake Bay:

KINGDOM ANIMALIA
  PHYLUM ARTHROPODA
    SUBPHYLUM CRUSTACEA
      CLASS MALACOSTRACA
        ORDER DECAPODA
          FAMILY PORTUNIDAE
            GENUS *CALLINECTES*
              SPECIES *CALLINECTES SAPIDUS*

Note that the names of taxa are capitalized. **Species** is identified by two names: the first name is capitalized, the second is not. Note also that those names are distinguished by either underlining them or setting them in *italics*.

Identification

The essence of species identification is the tracking of a specimen through the hierarchy of Linnaean categories. In some cases, you might recognize a familiar form such as a starfish and assign it to a phylum simply by searching an index of a reference text. When you encounter a species that simulates superficially the body plan of some other taxon, however, this method may prove dangerous. Several phyla of marine "worms," for example, look alike.

How to Use Keys

Practicing **systematists** have derived several aids that facilitate species identification. These aids are called **keys** because they systematically unlock the door to the unknown. Such keys are often based on conspicuous external morphology and, to some extent, on internal anatomy. Because many of the **characters** used in the keys may have no apparent

Figure 2.1 Phylogenetic tree of kingdom Animalia, modified from Margulis and Schwartz (1988). Fully capitalized names represent phyla; uppercase and lowercase taxa are subphyla or classes. Radiata includes the radially symmetrical organisms; bilateria the bilaterally symmetrical organisms (at some stage of the life cycle). These are further subdivided into those that lack a body cavity (acoelomates); those that have a cavity, but lack a true coelom (pseudocoelomates); and those that develop a true coelom (coelomates).

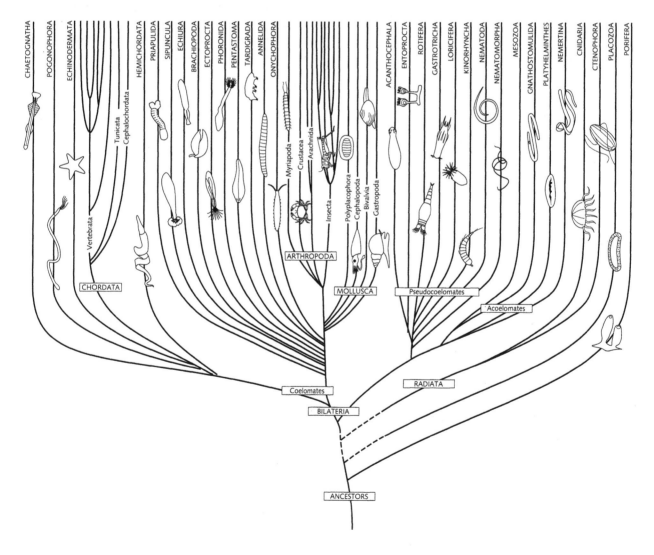

evolutionary or phylogenetic application, most taxonomic keys are "artificial." That is, they group species that are superficially similar in form, but not necessarily closely related phylogenetically.

The principal goal of a key is simple—to provide an easy and rapid route to identification. The simplest type of key is **dichotomous**; it provides you with an alternative choice in each step. Each statement should be judged (by you) to be true or false. The path in a dichotomous key is clearly indicated: if one alternative is false, then the next step is specified automatically. Following this route should result in the exclusion of a subset of taxa to which your specimen does not belong. Keys in which the alternative statements are placed together as **couplets** are relatively easy to use and, fortunately, the most commonly used form.

An example of a dichotomous key to three hardware fasteners follows:

1. a. Head slotted ------- screw
   b. Head smooth ------- 2 (that is, go to No. 2)

2. a. Tip pointed -------- nail
   b. Tip blunt ---------- bolt

EXERCISE 1

Using the Key to Marine Invertebrates

This exercise introduces you to techniques required to identify common groups of marine animals and to arrange these groups into an existing hierarchy of classification, as established by modern taxonomy. You will initially work with a generalized key that will enable you to identify an organism to higher taxa (from phylum to order).

The key included in this exercise is fundamentally dichotomous, but it deviates from that pattern (it becomes trichotomous) in certain situations when the final identification is made. To complete identification through family to species, you must use descriptive guides or keys that are applicable to the biogeographic province of your collection.

PROCEDURE

NOTE: If this is your first laboratory exercise that requires the use of microscopes, you may need instruction in their use (see Appendix I).

1. Select one of the specimens for study. Use dissecting trays for preserved specimens and keep

the specimens moist. Live organisms should be placed in clean, uncontaminated containers filled with seawater.

2. Use the Key to Marine Invertebrates included in this unit to identify the specimen's phylum, class, or order.
   NOTE: You may be asked to identify certain specimens to progressively lower taxa, perhaps even to species; they will be labeled accordingly. Ask the instructor to direct you to an appropriate reference guide and/or key.

3. Verify your identification.

   a. Verify that the location of collection corresponds to the known range of species distribution.

   b. Compare your specimen with published illustrations, descriptions, and reference specimens. Be aware that many species have similar appearances.

4. Record the specimen number, species and common names, and the names of higher taxa in your laboratory notebook.

5. Sketch the specimen and label the *conspicuous external parts that aid in the identification.* These are the characters used in the keys. Include a scale bar or indicate the size of the specimen.

6. Repeat steps 1–5 for each of the required number of specimens.

7. Research the following information that is available from text and reference material, and include it in your report:

   a. **Typical habitat** should be described as specifically as possible (for example, intertidal mud flat, benthic continental shelf, or planktonic estuary). This information can be obtained from reference guides and keys.

   b. **Characteristics of phylum and class** are the *natural features* (usually morphological) that distinguish one phylum from another, or one class from another within a phylum. Some of these characteristics cannot be observed without dissection, and therefore are not used in artificial keys. For example, sea squirts are members of the phylum Chordata because they possess a notochord, a structure that you can identify only if proper staining and microscopic examination are used. This information is available in textbooks. *Do not confuse natural characteristics with the artificial traits used to key an organism.*

   c. **References** are the guides used to make your identification. You need only include the author and date of your information source. The full citation should be placed in your laboratory notebook, or in the manual as you compile your bibliography.

Interpretation

1. In procedural step 5, you were advised to label your drawing to identify the *key characters* of each specimen. In procedural step 7, you were asked to list the *natural features* characteristic of the phylum and class of the specimen. Did you notice if any natural features were used in the artificial key? Was this more noticeable for certain taxa than for others? Provide examples in your answer.

2. Compare and contrast your identification of any of the worm groups with that of bivalve or gastropod mollusks. What, for example, are the features used in the key? Do you require live animals to make the identifications? Do you require the entire organism?

Consult your text or other suitable references for the following questions.

3. How many of the invertebrate phyla are considered to be of major importance? Define the highly subjective term "major" by considering, for example, whether importance is based on sheer numbers, conspicuous presence, size, or commercial value.

4. Examine a phylogenetic tree and make an effort to separate "lower" from "higher" invertebrates. The term "major" should not be used in this discussion.

5. Which of the invertebrate taxa observed in this exercise have the greatest probability of representation in the fossil record? Consider what is involved in fossilization, and which specimens possess structures most likely to be preserved as fossils.

KEY TO MACROSCOPIC COASTAL MARINE INVERTEBRATES

This key includes invertebrate forms that are frequently encountered in the shallow Atlantic and Pacific coastal zones of North and Central America, from boreal to tropical regions. The key emphasizes **benthic** organisms (living in, on, or near the ocean bottom). **Plankton** (passively floating or weakly swimming organisms) and **pelagic** (open water) forms are excluded. (Keys to plankton are included in Unit 8.)

This key is designed for use in this exercise as well as in all other field and laboratory exercises in which some form of identification is necessary.

You are initially asked to determine if the organism is permanently attached to the substratum **(sessile);** if it is, move to couplet 2 in the key. Next, you determine whether its growth form is erect (move to couplet 3) or flat and encrusting (your specimen should resemble one of the organisms pictured in the key). If the organism is **motile,** move to couplet 12 where you must decide if it is wormlike or non-wormlike. Refer to the accompanying figures as you work your way through the key.

You may encounter organisms that do not fit perfectly into the alternative descriptions. Choose the better of the two descriptions. In such instances it may be necessary to pursue both pathways to resolve the conflict. If you find that both choices of a couplet pair are false, retreat backward through the key until you discover where you have gone astray. In this way, you can avoid having to return to the beginning.

In a few instances, you will be asked to select the best description from among three choices (trichotomy). Examine each numerical step carefully to determine if you have two or three choices.

1 a.	Sessile form (firmly fixed to substratum)	2
b.	Motile organisms	12
2 a.	Growth form erect, upright, may be branched	3
b.	Growth form encrusting colonies that form low bumps, pits, or cushions	8
3 a.	Growth form is colonial (consisting of numerous individuals interconnected or conspicuously attached to each other); exceptions are solitary sponges (see, for example, couplet 6)	4
b.	Solitary animals, sometimes aggregated, but without structural connections	9
4 a.	Growth of small, delicate colonies with few to a large number of branches arising from a stolon (horizontal branch) or holdfast (point of attachment)	5
b.	Growth form in thick, soft, flexible lobes	6

5 a. Zooids (less than 5 mm), asexually produced individuals of a colony (5 cm or larger); single or multiple whorls of tentacles scattered on upright stalk; mouth at center of whorl of tentacles (can be withdrawn if disturbed); *colonial hydroids*

PHYLUM CNIDARIA
  CLASS HYDROZOA

b. Zooids (less than 5 mm) with single whorl of tentacles (can be withdrawn if disturbed); enclosed in tubular or boxlike shells; arranged in erect branching or fanlike growth pattern (5 cm or larger); *moss animals*

PHYLUM ECTOPROCTA (= BRYOZOA)

6 a. Body of individual or colony (size variable), spongy in nature with internal cavities and fibrous network of spicules (minute pointed hard bodies); surface porous, often with conspicuous, well-developed openings; growth form diverse; *sponges*

PHYLUM PORIFERA

b. Colonial form with zooids embedded in a common, calcareous (consisting of calcium carbonate, also called limestone), but somewhat flexible matrix   7

7 a. Individuals in colony polyplike (hollow cylinder with mouth opening at center; whorl of 8 tentacles surrounds mouth); polyps may be retractile. Colony large (to 60 cm); form diverse. Soft corals; a. *sea fans*, b. *sea whips*

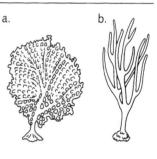

PHYLUM CNIDARIA
  CLASS ANTHOZOA
    SUBCLASS OCTOCORALLIA

b. Individuals embedded in variable size colony; saclike, without tentacles, frequently arranged in oval, circular, or starlike patterns with minute incurrent and excurrent openings (siphons). Colonial tunicates; *sea grapes, sea blubbers*

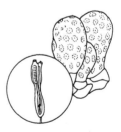

PHYLUM CHORDATA
  CLASS ASCIDIACEA

8 a. Encrusting colony spongy with internal cavities and fibrous network of spicules (minute pointed hard bodies); surface porous, often with conspicuous, well-developed openings; size and growth form diverse; *sponges*

b. Encrusting colonies of varying dimensions, flattened, stoloniferous, lacy, or sheetlike aggregations of minute zooids (less than 0.5 mm) with tentacles (can be withdrawn if disturbed); encased in individual boxlike or tubular calcareous (consisting of calcium carbonate, also called limestone) shells; *moss animals*

PHYLUM ECTOPROCTA (= BRYOZOA)

c. Zooids embedded in gelatinous or rubbery matrix; of varying size; encrusting or partly erect; more or less amorphous, lobular, or hemispherical; zooids often in starlike patterns; *colonial tunicates* (see also couplet 7b)

PHYLUM CHORDATA
    CLASS ASCIDIACEA

d. Colonies generally of short polyplike individuals (hollow cylinder with central mouth surrounded by variable number of tentacles at one end) about 5 mm in diameter inter-connected by living tissue that covers calcareous skeleton; individuality of polyps lacking in some; tissue forms meandering rows on skeleton; growth form of colony diverse; a. antlerlike, b. hemispherical, platelike; *colonial stony corals*

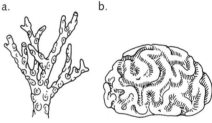

PHYLUM CNIDARIA
    CLASS ANTHOZOA
        ORDER SCLERACTINIA

9 a. Calcareous (calcium carbonate) forms; conical shape or bivalved (2 shells)　　　　10

b. Polyplike (hollow cylinder with mouth opening at center of whorl of tentacles), or with calcareous plates at top of stalk, or saclike with two conspicuous openings (siphons)　　　　11

10 a. Conical form consisting of closely interlocking calcareous plates; base diameter to about 2 cm; opening in living animal sealed by two, bipartite (two-part) valves; *acorn barnacles*

PHYLUM ARTHROPODA
    SUBPHYLUM CRUSTACEA
        CLASS MAXILLOPODA
            SUBCLASS CIRRIPEDIA

b. Clamlike, with two shells (valves); lower one cemented to substratum *(oyster)*; or organism is attached by tuft of long tough filaments (byssus threads); *(mussel)*

PHYLUM MOLLUSCA
  CLASS BIVALVIA (= PELECYPODA)

11 a. Polyplike; cylindrical column (few millimeters to several centimeters in diameter) attached to substratum by pedal disc; oral surface with simple (one ring) or complex (multiple rings) whorl of tentacles encircling a central mouth; *sea anemones*

PHYLUM CNIDARIA
  CLASS ANTHOZOA
    ORDER ACTINIARIA

b. Stalked form; stalk scaly and/or rubbery; body compressed laterally (flattened side to side), and armored with calcareous (calcium carbonate) plates; *gooseneck barnacles*

PHYLUM ARTHROPODA
  SUBPHYLUM CRUSTACEA
    CLASS MAXILLOPODA
      SUBCLASS CIRRIPEDIA

c. Saclike, frequently encrusted with foreign debris; with two siphons [tubes for water transport into (incurrent) and out of (excurrent) the body] at upper end; size variable, to 5 cm; *solitary sea squirts*

PHYLUM CHORDATA
  CLASS ASCIDIACEA

12 a. Wormlike, elongated, cylindrical, threadlike or ribbonlike animals, as well as shorter, stouter, cucumberlike or peanut-shaped forms; soft disclike or filmlike types   **13**

b. Not wormlike as described above; some lack exoskeletons (external skeleton); others have more or less rigid exoskeletons; resemble snails, clams, crabs, shrimps, lobsters, insects   **20**

13 a. Body distinctly segmented (divided into multiple, repetitive parts), or regularly ringed, sometimes with papillae (small nipplelike projections) or spines, or faintly annulated without tube feet (flexible, suckerlike projections)   **14**

b. Body not segmented, or if faintly annulated, then with tube feet   **16**

14 a. Body with 20–40 rings of papillae or spines 15

　　b. Segmentation usually distinct, or not with papillae or spines; conspicuous parapodia (lateral body appendages), simple or lobed, with chitinous (horny) setae (bristles) (singular, seta), and cirri (soft, finger-like projections that serve as gills); head usually with tentacles, or other appendages; size variable, 1 to 20 cm. Bristle worms; a. *clam worms*, b. *crested worms*, c. *fan worms*, d. *scale worms*

PHYLUM ANNELIDA
　　CLASS POLYCHAETA

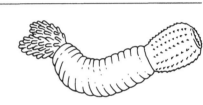

15 a. With 30–40 rings of papillae; posterior end with terminal clusters of contractile appendages; anterior end with completely retractile, bulbous spiny proboscis (flexible, tubular process of the oral region) and short collar; size variable, 1 to 10 cm; *priapulid (club) worms*

PHYLUM PRIAPULA (= PRIAPULIDA)

　　b. With about 22 spinous rings; anterior end with nonretractile (cannot be drawn into mouth) spoon-shaped proboscis (flexible tubular process); size to 20 cm; *spoon worms*

PHYLUM ECHIURA (= ECHIURIDA)

16 a. Body usually more or less sluglike with a solelike creeping foot; head usually with one or more pairs of tentacular appendages; back with lateral folds (mantle) and soft, clublike dorsal appendages in rows or scattered, or with a cluster of retractile gills posteriorly; with or without an internal shell; to 15 cm; *nudibranchs, sea hares*

PHYLUM MOLLUSCA
　　CLASS GASTROPODA

　　b. Not as above 17

17 a. Anterior end with a circlet of retractile tentacles or lobes and with body separated into distinct regions (regionated) or not, or with a nonretractile (cannot be withdrawn into mouth) proboscis or single long tentacle and with body regionated 18

　　b. Anterior end without a proboscis or circlet of tentacles or, if with a retractile proboscis, with body regions ill-defined or a short caudal cirrus (fingerlike projection) 19

18 a. With a nonretractile, stalked proboscis, short collar, and trunk (sometimes regionated); to 3 cm; *acorn worms*

PHYLUM HEMICHORDATA

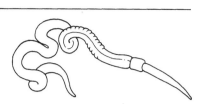

b. With a retractile proboscis with terminal mouth more or less surrounded by tentacles or lobes; to 10 cm; *peanut worms*

PHYLUM SIPUNCULA (= SIPUNCULIDA)

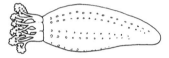

c. With a retractile circlet of tentacles; skin smooth, false appearance of segmentation, or wrinkled, usually with minute, embedded calcareous (calcium carbonate) deposits; with or without tube feet; mouth anterior; anus posterior; to 30 cm; *sea cucumbers*

PHYLUM ECHINODERMATA
CLASS HOLOTHUROIDEA

19 a. Body usually depressed dorso-ventrally (flat in cross section); often disc-shaped; gut saclike or with radiating diverticula (outpocketings of gut seen only in transparent forms with microscope); mouth ventral, no anus; to 15 mm; *flatworms*

PHYLUM PLATYHELMINTHES
CLASS TURBELLARIA

b. Body depressed, or oval in cross section; usually greatly elongated; with retractile proboscis; anus terminal but obscure; to 6 cm; *ribbon worms*

PHYLUM NEMERTEA

c. Body usually round in cross section; mouth surrounded by circlets of minute papillae; anus terminal; to 10 mm; *roundworms*

PHYLUM NEMATODA

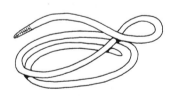

20 a. With 1, 2, or 8 calcareous (calcium carbonate) shells 21

b. With or without a calcareous or chitinous (horny) exoskeleton but not as above 23

21 a. With two shells (valves); anchored by byssus threads, or shell cemented to substratum (see couplet 10), or burrowing forms; *bivalves, clams*

PHYLUM MOLLUSCA
   CLASS BIVALVIA (= PELECYPODA)

b. With 8 dorsal (upper-surface) plates in a row; head and tail shells semicircular in shape, others rectangular or winged; body oval with fleshy marginal girdle; *chitons*

PHYLUM MOLLUSCA
   CLASS POLYPLACOPHORA

c. With one shell   22

22 a. Shell uncoiled, conical, or cuplike; a. *limpets,* b. *slipper* or *boat shells*

PHYLUM MOLLUSCA
   CLASS GASTROPODA

a.   b.

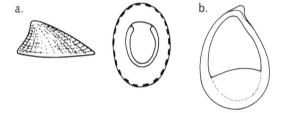

b. Shell distinctly coiled; *snails*

PHYLUM MOLLUSCA
   CLASS GASTROPODA

23 a. Forms with radial symmetry; with skeleton of calcareous (calcium carbonate) ossicles (small calcareous plates) either fused together or embedded in skin; body parts usually in fives or multiples of five   24

b. Forms with bilateral symmetry   27

24 a. Distinct arms (extensions from central body) present, radiate from body   25

b. Arms absent   26

25 a. Five arms sharply marked off from central disk; tube feet reduced in size, papillalike (nipplelike, rather than with flattened bottom), in pores on ventral (lower) surface of arm; *serpent (brittle) stars*

PHYLUM ECHINODERMATA
  CLASS OPHIUROIDEA

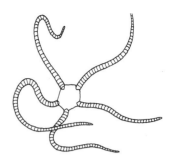

b. Five or more arms merging gradually with central disc; tube feet in 2 or 4 rows in grooves on ventral surface of arms; *seastars*

PHYLUM ECHINODERMATA
  CLASS ASTEROIDEA

26 a. Body a globular or disclike test of close-fitting plates, with numerous movable spines; a. *sea urchins,* b. *sand dollars*

PHYLUM ECHINODERMATA
  CLASS ECHINOIDEA

a. b.

b. Body cylindrical with soft, tough, leathery wall with minute ossicles embedded in skin; tube feet in 5 rows along body; spines absent; oral tentacles present (see couplet 18c); *sea cucumbers*

PHYLUM ECHINODERMATA
  CLASS HOLOTHUROIDEA

27 a. Segmented exoskeleton and jointed legs; body subdivided into regions; diverse forms (millipedes, centipedes, spiders, insects, crustaceans)   28

PHYLUM ARTHROPODA

b. Saclike animals; with 2 well-developed, complex eyes; with 10 sucker-bearing arms surrounding mouth; *octopus*

PHYLUM MOLLUSCA
  CLASS CEPHALOPODA

28. a. Antennae (elongated, movable, segmented organs on head) present   29

    b. Antennae absent   30

29 a. Two pairs of antennae present; mandibulate (holding or biting mouthparts); usually at least 5 pairs of appendages (legs) on thorax (second region of body); body usually distinctly segmented; crustaceans   31

PHYLUM ARTHROPODA
   SUBPHYLUM CRUSTACEA

    b. One pair of antennae present; a. *millipedes*, b. *centipedes*, c. *insects*

PHYLUM ARTHROPODA
   SUBPHYLUM UNIRAMIA

a.

b.

c.

30 a. Large animals with two distinct body regions—a large horseshoe-shaped anterior region (prosoma) and a posterior region (opisthosoma) armed with 6 small spines on each side; opisthosoma contains leaflike appendages bearing gills; abdomen terminates in long spikelike telson; *horseshoe crabs*

PHYLUM ARTHROPODA
   SUBPHYLUM CHELICERIFORMES
     CLASS CHELICERATA
       ORDER XIPHOSURA

    b. Small, spiderlike animals with reduced prosoma; 4 pairs of legs; head with terminal proboscis, and 1 or 2 pairs of specialized appendages; *sea spiders*

PHYLUM ARTHROPODA
   SUBPHYLUM CHELICERIFORMES
     CLASS PYCNOGONIDA

NOTE: All forms beyond this point in the key are members of the subphylum Crustacea, class Malacostraca. Lower taxa are designated at the appropriate locations in the key.

31 a. Carapace (shell enclosing all or part of head and thorax of body) present   32

    b. Carapace lacking   39

c. Carapace short, leaving last 4 or 5 thoracic segments exposed; eyes stalked; abdomen longer than thorax, often flattened; chelae raptorial (seize prey similar to those of mantid insect); *mantis shrimp*

ORDER STOMATOPODA

32 a. Shrimplike or lobsterlike with well-developed abdomen and caudal (tail) fan (uropods and telson)   33

    b. Crablike or egg-shaped with abdomen and its appendages soft and degenerate, or curled under body   36

33 a. Thoracic legs 1–3 chelate   34

    b. One or 2 pairs of thoracic legs chelate; never the third pair   35

34 a. Chelae (singular, chela) not strongly developed; *penaeid shrimp*

ORDER DECAPODA
    INFRAORDER DENDROBRANCHIATA

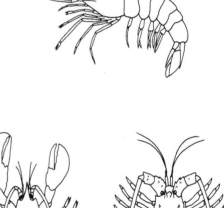

    b. First pair of chelae strongly developed (note exception in spiny lobsters, the palinurids; chelae not strongly developed); *chelate lobsters, spiny lobsters*

ORDER DECAPODA
    INFRAORDER ASTACIDEA
      INFRAORDER PALINURA

35 a. Chelae on first pair of legs strongly developed; *ghost shrimp*

ORDER DECAPODA
  INFRAORDER THALASSINOIDEA

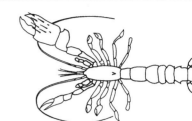

  b. Chelae on first pair of legs not greatly developed, or if so,
     then eyes nearly covered by carapace; "broken back" profile
     often evident in abdomen; *caridean shrimp*

ORDER DECAPODA
  INFRAORDER CARIDEA

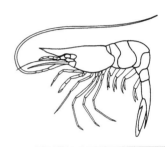

36 a. Body egg-shaped; none of legs chelate; *mole crabs*

ORDER DECAPODA
  INFRAORDER ANOMURA
    FAMILY HIPPIDAE

  b. Not as above   37

37 a. Crablike in front part of body; abdomen soft and degenerate;
       somewhat curled and with reduced appendages; animal lives
       in gastropod mollusk shell; *hermit crabs*

ORDER DECAPODA
  INFRAORDER ANOMURA
    FAMILY PAGURIDAE
      FAMILY DIOGENIDAE

  b. Abdomen tightly flexed against underside of body   38

38 a. Last pair of walking legs (pereopods) (fourth pair behind the
       chelae) greatly reduced or apparently absent; *porcelain crabs*

ORDER DECAPODA
  INFRAORDER ANOMURA
    FAMILY PORCELLANIDAE

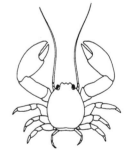

b. Last pair of pereopods about the same size as others; 5 pairs of appendages conspicuous; *true crabs*

ORDER DECAPODA
INFRAORDER BRACHYURA

39 a. Body compressed side to side; 3 pairs of uropods; *scuds, beach hoppers*, and *skeleton shrimp (caprellids)*

ORDER AMPHIPODA

b. Body cylindrical or depressed top to bottom; one pair of uropods; *sow bugs*

ORDER ISOPODA

Benthic Invertebrates

OBJECTIVES

After completing this unit, you will be able to

- Describe salient features of conspicuous or numerically abundant benthic invertebrate animals;

- Relate observed morphology of the animals to their particular habitats and niches (functional morphology and survival strategies), with emphasis on feeding; and

- Recognize planktonic larval forms of benthic animals.

INTRODUCTION

Benthos is a word of Greek origin that means depths of the sea. It also refers to marine organisms that live on, in, or near the bottom of a body of water. Animals living *on* the bottom, such as sponges, sea anemones, and tunicates, are called **epifauna.** Animals living *in* the sediment, such as bivalve mollusks and polychaete worms, are called **infauna.**

Benthic organisms can be classified according to size, designated by prefix, such as **macrobenthos** (larger than 0.5 millimeter = mm). **Meiobenthos** are smaller than 0.5 mm, but large enough to be retained by a sieve with a mesh size of 63 micrometers = μm = 0.063 mm. **Microbenthos** (unicellular protists and bacteria) are less than 50 μm in size.

Bottom-dwelling organisms also can be classified according to their ability **(motile)** or inability **(sessile)** to move. Sessile benthos—for example, barnacles and sea squirts—are attached to stable surfaces on the sea floor.

A naturally occurring assemblage of interacting populations of plants and animals that occupy the same benthic area is a **benthic community.** The numerical abundance of any given population in the community is influenced by mortality and recruitment, and by immigration and emigration of individuals. Survival and reproductive success of in-

dividuals and populations are affected by abiotic factors such as available light, temperature, dissolved oxygen, salinity, and other dissolved minerals.

The **habitat** is the place in which an organism (or an entire community) lives; it includes biotic and abiotic factors. The distribution of certain organisms within the community is often controlled by microdifferences in abiotic factors, which create what are considered to be microhabitats.

Each species within a habitat occupies a **niche**—that is, a certain function within the community. This niche is defined by interspecific (between individuals of different species) and intraspecific (between individuals of the same species) relationships, abiotic requirements, and the organisms' influence on the abiotic environment.

Conspicuous infaunal and epifaunal invertebrate animals are described in this unit. They represent several major taxa, and different anatomical body plans. The animals' structural designs are related to habitat and niche, and represent different solutions to common problems of survival. The organisms were selected on the basis of their common and ubiquitous occurrence in coastal marine environments. They are presented according to the feeding strategy that they represent: predators/scavengers, grazers, deposit feeders, and suspension feeders. Because these benthic invertebrates produce larval stages that are planktonic (see Fig. 8.1 in Unit 8), their contributions to the meroplankton are described.

GENERAL PROCEDURE

1. Obtain the designated living specimens (handle them with care). Place them in appropriate-size *clean* containers filled with seawater from the aquarium containing the specimens.
   NOTE: If you are using preserved specimens, take appropriate safety precautions (see Unit 2) and use containers designated for use with preservatives.

2. Under no circumstances should living animals be dissected or mutilated. If they are to be dissected, they must first be anesthetized. Submersion in a $MgCl_2$ solution isosmotic to seawater is an effective anesthetic procedure, but takes time. Plan ahead.

3. Study the specimens according to the specific directions in each exercise.

4. You will find questions throughout the descriptions of representative specimens that are designed to sharpen your observational skills. You should attempt to answer these questions in your notebook as you work.

5. Refer to the accompanying figures and available references for assistance. A checklist of the information for which you may be held responsible follows (items in **bold type** must be observed in the laboratory):

   a. Identification: common name, scientific name.

   b. Specific habitat: epifauna, infauna, pelagic, sessile, motile?

   c. **External features** (drawing): complete with legend, labels, and indication of size or magnification.

   d. **Feeding mechanism:** What and how does it eat? What structures are used? Is it a predator, grazer, scavenger, suspension or deposit feeder?

   e. **Respiratory gas exchange:** organs involved, and mechanism, in relation to your drawings.

   f. **Internal anatomy** (drawing): complete as in part (c).

   g. **Unique traits:** size, sex, color, maturity, anomalies, ovigerous (egg-bearing), and behavior may be illustrated and described on the report form. If there is more than one specimen of the species, note any observable variations.

   h. Reproduction: Are sexes separate (dioecious), or are testes and ovaries found in the same individual (monoecious), or is asexual reproduction evident? Can you see **secondary sex characters?**

   i. Development: metamorphic, **planktonic larvae?**

6. You may be required to perform some of the exercises from the following General Interpretation section. *Work the exercises only after the laboratory observations are completed.*

General Interpretation

1. Summarize the observations you made during this exercise. Construct a table and place the representative organisms in the first column on the left. In sequential columns across the page, list the principal features that were observed (select from the list in procedural step 5). Then fill in the empty boxes for each organism, using key descriptive words or short phrases.

2. Based upon your table from question 1, provide the following additional information and/or answer the following questions:

   a. Construct a food chain (or food web) that includes the observed cast of characters. Indicate the trophic level of each.

   b. Which of your specimens possess exposed gills? Defend your answer.

   c. Which of your specimens depend upon mucous for feeding? For locomotion?

3. Can you assume that the specimens you observed are *absolutely* representative of their respective taxa? For example, if you examined a particular crab, is it safe to apply your observations to other species of crab? Defend your answer, using examples. Consult your textbook, if necessary.

4. The Key to Marine Invertebrates in Unit 2 uses the following descriptions of body form: wormlike, shelled, polyplike. Place your representatives in these categories. Is the taxon (phylum, class, and so on) of your representative the only taxon that fits into those categories? If not, name at least one other taxon that fits that description, and then indicate if it matches the other characteristics described for your representative in this exercise.

EXERCISE 1
Predator/Scavenger: Crab

CRAB
  PHYLUM ARTHROPODA
    SUBPHYLUM CRUSTACEA
      CLASS MALACOSTRACA
        ORDER DECAPODA
          INFRAORDER BRACHYURA

Crabs are perhaps the most diverse and successful groups of the subphylum Crustacea. The 4500 described species represent more than half of the order Decapoda (the ten-legged crustaceans), which also includes lobsters and crayfish.

Crabs are epibenthic omnivores (organisms that consume a variety of foods). They eat dead organisms, but are also known to hunt, stalk, and ambush live prey. Some feed on algae and particulate matter.

Brachyuran crabs are quite diverse in morphology, with their variations being closely related to their habitat and niche. The following description applies to the more commonly occurring (and commercially available) species of brachyuran crabs, such as the eastern rock crab *(Cancer irroratus)* and green crab *(Carcinus maenas)* of the New England coast, the blue crab *(Callinectes sapidus)* of the Middle Atlantic through Gulf of Mexico region, and the western rock crab *(Cancer antennarius)* and striped shore crab *(Pachygrapsus crassipes)* of the California coast.

PROCEDURE

External Features

1. Examine the external morphology of available specimens.

   Brachyuran crabs, as the name implies (*Brachy* = short, *ura* = tail), have shortened bodies; most are wider than they are long. Variations in body depth and shape are related to habitat.

   a. Sketch and describe the body shape of your specimen, and try to relate it to the natural habitat of the crab.

   b. Measure the length, width, and body depth of your specimen. Calculate ratios of width to length, and depth to length. Values close to one indicate equal dimensions. Is your specimen longer than it is wide? Relatively flat or thick?

2. Observe the dorsal surface (Figs. 3.1, 3.2).

   The head and thorax regions (**cephalothorax**) of the body are enclosed by a broad, flat **carapace.** Note the presence and location of any conspicuous ridges, spines, or setation (hairy surface). The ridges delineate regions that correspond to internal morphology. For example, the gastric mill and the heart lie directly below the gastric and cardiac regions, respectively.

   c. Sketch the dorsal carapace and describe its color and any distinguishing features. Do these traits appear to have any adaptive advantage?

3. Turn the crab on its back, and examine the ventral side. If the crab is alive, you may require assistance. Ask your instructor for the proper way to handle a live crab. Unlike the lobster, the crab possesses an **abdomen** that is reduced in size and musculature, and is folded (flexed) under the cephalothorax (Figs. 3.1b and 3.2b,c). The sex of the animal can be determined by the shape of the abdomen, which fits snugly in a depression in the skeletal plates that make up the **thoracic sternum** (Figs. 3.1b,c and 3.2b,c). In most crab species, the abdomen of the male is acutely triangular (Fig. 3.1b), and some abdominal segments may be fused. In *Callinectes* spp., the male abdomen has an inverted T-shape (Fig. 3.2b). In mature females, the abdomen is usually broadly triangular to semicircular in shape, and the individual segments are conspicuous (Figs. 3.1c and 3.2c). In some species—for example, *Callinectes* spp—the immature females have a triangular abdomen (Fig. 3.2d) that changes to the broad shape (Fig. 3.2c) at maturity.

4. Carefully lift the abdomen and observe the paired appendages (Figs. 3.1 and 3.2). The male has two pairs of pleopods that are modified to form **gonopods** used in copulation (Figs. 3.1c and 3.2b). At the time of mating, the first pair are inserted into the **gonopores** of the female (paired openings on the female's sternum that are normally covered by her abdomen). The mature female has four pairs of **biramous** (two branches) and heavily setated **pleopods** (Figs. 3.1d and 3.2c). Eggs are cemented to these setae at the time of spawning. While the female is **ovigerous** (bearing eggs), the abdomen is extended to accommodate the large egg mass.

   d. Sketch the ventral aspect of your crab and indicate its sex.

5. Examine the legs. The crab has five pairs of legs (**pereopods**)—hence the term "decapod." The first pair are pincerlike claws called **chelae** (plural) or **chelipeds** (Figs. 3.1b and 3.2a). The right and left chelipeds often differ in size and **dentition** (the types of teeth and spines on the edges of the fingers of the claw).

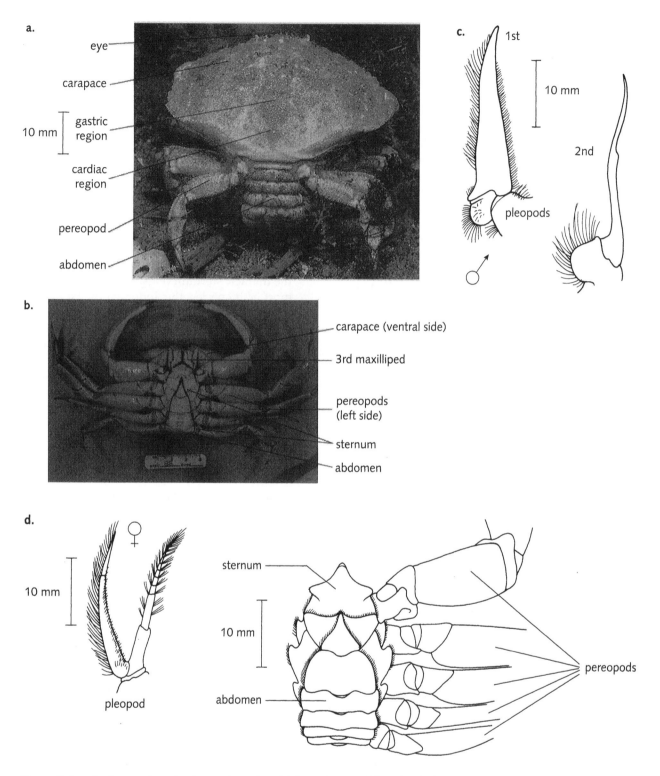

a.

eye

carapace

10 mm | gastric region

cardiac region

pereopod

abdomen

c.

1st

10 mm

2nd

pleopods

b.

carapace (ventral side)

3rd maxilliped

pereopods (left side)

sternum

abdomen

d.

♀

10 mm

pleopod

sternum

10 mm

abdomen

pereopods

Figure 3.1. Eastern rock crab, *Cancer irroratus.* (a) Dorsal view of female, abdomen partially extended. (b) Ventral view of male with abdomen in flexed position. (c) First and second pleopods (gonopods). Second gonopod pumps sperm through hollow shaft of first gonopod. (d) Ventral view of female, legs of left side only shown. One of four pairs of biramous pleopods shown at left.

a.

fifth pereopod

cardiac region

gastric region

carapace

eye in orbit

cheliped

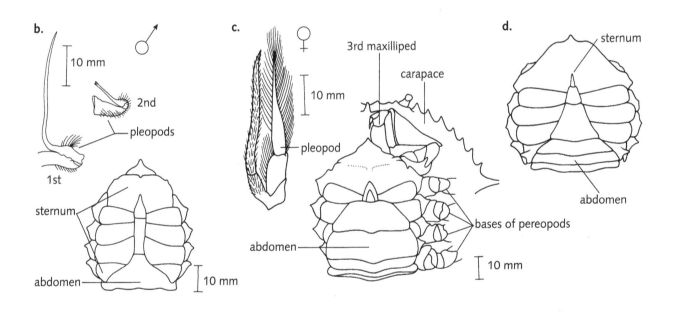

b.

10 mm

2nd

pleopods

1st

sternum

abdomen

10 mm

c.

10 mm

pleopod

3rd maxilliped

carapace

abdomen

bases of pereopods

10 mm

d.

sternum

abdomen

Figure 3.2 Chesapeake Bay blue crab, *Callinectes sapidus*. (a) Dorsal view. Paddle-shaped fifth pereopods are obvious. (b) Ventral view of male with inverted T- shaped abdomen. First and second pleopods (gonopods) shown to the left. Second gonopod pumps sperm through hollow shaft in first gonopod. (c) Ventral view of mature female with broadly rounded abdomen. One of four pairs of biramous pleopods shown at left. (d) Triangular abdomen of immature female.

e. Gently probe your specimen with a pencil. What does it do when threatened? Is it aggressive, or does it avoid you by backing away (and/or digging into the aquarium gravel bed)? Does it grab the pencil? Does it release the pencil quickly? Did it make any indentation on the pencil?

f. Sketch the claws of your specimen and describe the dentition (finer and more serrated teeth for cutting, or fewer and blunt for crushing).

The other four pairs of legs may be similar in size and shape to one another, as in *Cancer* sp. (Fig. 3.1b), or the fifth pair may be modified for swimming (Fig. 3.2a). In that case, the terminal segments will be flattened and expanded as paddles. This feature is typical of most members of the family Portunidae (swimming crabs).

g. Sketch the last pair of appendages in your crab. Do they fit the description of a portunid crab?

6. Study the **frontal** region of the crab (Figs. 3.1 and 3.2). Note the position of the **compound eyes** on **eyestalks** lying within orbits. Examine the structure of the **antennules** and **antennae** (Fig. 3.3a,b), which contain chemical and tactile (touch) receptors.

h. Describe the response when you touch these appendages. Can they be retracted into a slot in the carapace?

7. Examine the mouthparts. There are six pairs of mouthparts: a pair of **mandibles,** two pairs of **maxillae,** and three pairs of **maxillipeds** (Fig. 3.3).
NOTE: Complete morphology of mouthparts can be observed only if they are removed from a preserved specimen. If time permits, and if preserved specimens are available, you may be asked to remove the mouthparts and study their comparative morphology.

The most conspicuous pair are the platelike **third maxillipeds** (Figs. 3.1b, 3.2c, and 3.3c), which cover the other mouthparts (Fig. 3.3d–h)

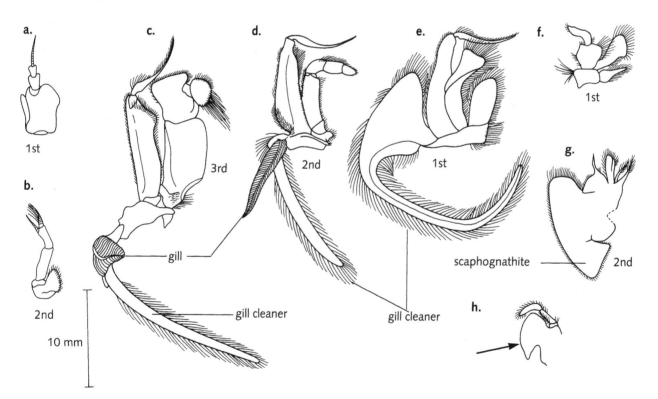

Figure 3.3 Appendages of *Cancer irroratus.* (a) First antenna. (b) Second antenna (antennule). (c–h) Mouthparts: (c, d, e) Maxillipeds. Setated gill cleaner extensions are situated in the gill chamber, and are responsible for circulating the water and cleaning the gills. Note also presence of gills on first and second maxillipeds. (f and g) Maxillae. Second maxilla contains the scaphognathite (gill bailer) that generates the ventilatory current through the gill chamber. (h) Mandibles (arrow points to grinding surface).

when at rest. The first and second maxillipeds are slightly smaller than the third pair. All of the maxillipeds possess secondary appendages that are involved in gill maintenance (see the discussion of ventilation and internal anatomy later in this exercise). During feeding, the third maxillipeds separate and expose the other mouthparts. The **maxillae** (Fig. 3.3f,g) are relatively small and not easy to distinguish from parts of the maxillipeds. The second maxilla has a secondary appendage that is involved in ventilation of the gill chamber. The **mandibles** (Fig. 3.3h) are the innermost mouthparts. They are compact and hardened, and they may contain toothed ridges that grind and masticate the prey.

8. Try feeding your specimen.

   i. Do you observe any increased activity of the mouthparts?

   j. Describe the feeding behavior. Can you determine which appendages are used, and how?

Locomotion

1. Allow the crab to move freely in a large aquarium, or place it on the floor.

   k. Describe the direction of travel. Is it straight ahead, or sideways?

   l. Describe how it carries the chelipeds when it is traveling.

2. If you have a portunid crab, entice it to swim.

   m. Describe the position and motion of the appendages as it swims.

Ventilation (Gas Exchange)

1. Allow your specimen to sit undisturbed in a bowl of seawater for 5–10 min. The water should just cover the crab.

   n. Look carefully at the surface of the water above the frontal region of the crab. Can you detect a rapid flow of water? If not, try placing a few drops of a diluted suspension of India ink (in seawater) near the bases of the legs where they join the thorax. Use a long Pasteur pipet and keep your distance from the claws!

   o. Can you determine the source of the current flow? Look for water entering the branchial chamber near the bases of the legs and flow-

ing out through narrow openings (**excurrent apertures**) on each side of the mouth. The water passes over the gills where gaseous exchange occurs. The current is produced by a specialized part (**scaphognathite**) of the second maxilla (Fig. 3.3g). The functional anatomy of ventilation and gas exchange is described in the dissection that follows.

Internal Anatomy

1. When you finish the study of external anatomy and behavior, obtain a specimen that has been prepared for dissection.

2. Remove the dorsal carapace. With strong scissors, make a shallow cut along the posterior margin on a line approximately 5 mm inside the posterior edge of the carapace. Continue this cut along each side and along the anterior margin just behind the orbits of the eyes. Return to the posterior margin, and carefully lift and separate the carapace from the underlying pigmented tissue (**hypodermis**). Strip away the hypodermis to reveal the underlying organs (Fig. 3.4). A large stomach (**gastric mill**) and associated median **gastric muscles** lie in the anterior part of the visceral cavity, in the midline. They support the mill and assist in the churning action. The large **heart** is immediately posterior to the gastric mill. **Hemolymph** enters the heart through **ostia,** two pairs of prominent valves on the dorsal surface. Posterolateral to the heart are **pericardial sacs** that serve as reservoirs for the blood (hemolymph). **Ovaries** (in females) or **testes** and **vasa deferentia** (in males) are H-shaped organs that lie on either side of the gastric mill and on top of lobes of the **digestive gland.** Gonadal tissue differs in texture and color from the digestive gland. In females (Fig. 3.4a), right and left branches of the ovary join immediately posterior to the stomach. A pair of oval-shaped **spermathecae** (seminal receptacles) extend ventrally to open externally through the **gonopores.** The spermathecae can be viewed only after the gonads and hepatopancreas are removed (later in the dissection). In males (Fig. 3.4b), a pair of cream to white colored **vasa deferentia,** where **spermatophores** (egg-shaped packets of sperm) are produced, extend postero-medially from the narrow, coiled testes. Vasa deferentia open externally into the **penes,** which insert in the first pair of gonopods on the abdomen (see Figs. 3.1c and 3.2b).

3. Remove a small portion of the vas deferens or ovary, and make a wet mount for microscopic examination.

   p. Draw or describe what you see.

   q. If your microscope has an ocular micrometer, you may be able to measure spermatophores or ova. See Appendix I for further details.

4. Remove the heart, gonad, and digestive gland, and trace the digestive tract. The short esophagus opens into the gastric mill. Digestive diverticulae (fingerlike pouches), or **caecae** (Fig. 3.4b), discharge enzymes into both the gastric mill and the intestine, which extends into the abdomen.

5. Remove the gastric mill by cutting the esophageal and intestinal connections. Place it in a

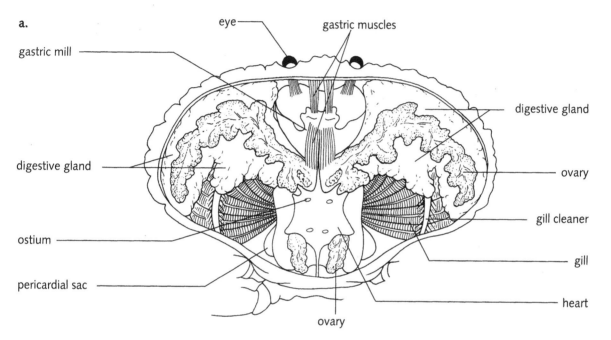

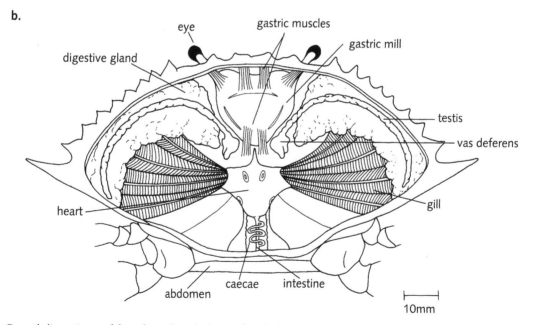

Figure 3.4 Dorsal dissections of female rock crab (a), and male blue crab (b). Top of carapace and underlying tissue layer removed to reveal conspicuous internal organs.

bowl of seawater, open the dorsal wall of the mill, and flush out any contents.

    r. Can you determine what the crab has eaten recently? Are whole organisms present, or are the remains largely fragments of organisms?

6. Examine the internal dentition of the mill. You may want to remove the teeth and examine them under a dissection microscope. The anterior lateral teeth are usually large and blunt for crushing and chewing. The median and posterior teeth are finer and sharp for accessory tearing and straining. Filter plates on the floor of the mill prevent rough particles from passing into the intestine or digestive gland.

7. Examine the branchial chamber. You may have to cut through the lateral wall of the carapace on one side to gain access to the bases of the gills. The **gills** lie on a shelf of the **endoskeleton** and converge to an apex (Fig. 3.4). Lying across the upper surface of the gills in each **branchial (gill) chamber** is the **gill cleaner,** a long setated (hairy) extension of the first maxilliped (Fig. 3.3c). The other maxillipeds (Fig. 3.3d,e) have similar extensions that lie beneath the gills. These three appendages circulate water within the branchial chamber and remove small organisms and debris from the surface of the gills. Note that the first and second maxillipeds also contain a gill attached to the base of the gill cleaner. A paddlelike extension (**scaphognathite,** or **gill bailer**) of the second maxilla propels water through the branchial chamber (Fig. 3.3g). The scaphognathite is located in the anterior part of the chamber, near the excurrent apertures.

    s. How many gills does the crab have? Are they attached to the endoskeleton or to the bases of mouthparts or pereopods? Are all of the gills the same size?

8. Remove a gill and examine the lamellar arrangement. The gills are **phyllobranchiate** (leaflike) in that they are composed of stacks of platelike **lamellae.** Each lamella is a thin, flattened chamber through which hemolymph flows. Exchange of gases occurs through thin layers of chitin and epithelium.

    t. Is the gill clean? Is there any evidence of parasitic infestation (organisms growing on, or embedded in, the gill tissue)?

9. Remove a portion of the membranous floor (part of the endoskeleton) of the branchial chamber to expose the **muscles** of the pereopods. These muscles constitute the principal source of crabmeat for human consumption.

10. Break open one of the chelipeds to reveal the muscles in the claw apparatus. Note that the muscles are attached to the internal surface of the larger segment of the claw, and to a large flat plate (**apodeme**) extending from the finger of the claw. Force of closure is related to the size and shape of the muscle and apodeme.

    u. Compare the bulk of the musculature in the body with that in the chela.

Larval Stages and Development

Examine prepared slides of the larval stages of a brachyuran crab.

Larval development in Brachyura is metamorphic (see Fig. 8.1 in Unit 8, and Key E, 8a and 10b, in the Keys to Plankton). The first postembryonic stage (hatched from egg) is a **zoea,** a free-swimming, planktonic stage. The zoea typically has a short helmet-shaped carapace with large rostral spikes that project in front. In addition, a pair of lateral spines may be present. The eyes are large but not set on stalks.

Several molts and metamorphoses produce the next form, the **megalopa.** This form resembles a crab more closely, but it is somewhat longer than broad. The spines of the carapace are lost and the body is somewhat flattened dorso-ventrally.

After about a month, the megalopa molts and metamorphoses into the first crab stage. This form settles out of the plankton and takes up a benthic existence.

Interpretation

1. Crabs are shelled invertebrates. How do they grow? Compare their growth with invertebrates that lack an exoskeleton. Consult your text or other reference material.

2. Crabs are typically messy eaters, and routinely ingest shell, mud, and sand, along with their intended meal. What structures prevent such "roughage" from passing into the intestine?

3. Did your specimen reveal warning or camouflage coloration? Explain.

4. Would you say that crabs have good eyesight? Defend your answer based upon your observations.

5. Did your specimen lack any appendages? If so, will it be replaced? If replacement occurs, when and how will the new appendages develop? Consult appropriate references.

6. Crabs have planktonic larvae. What are some advantages and disadvantages of planktonic development?

EXERCISE 2

Predator: Seastar

SEASTAR
PHYLUM ECHINODERMATA
CLASS ASTEROIDEA

Approximately 1600 species of asteroids (seastars) are distributed throughout the oceans. Some are voracious carnivores and feed by everting the stomach on oysters, barnacles, clams, and other stationary marine animals. Others feed on detritus, or obtain food from ingested mud.

Several common species of seastar (*Asterias forbesi*, North Atlantic; *Echinaster spinulosus*, Southeast Atlantic and Gulf of Mexico; *Pisaster giganteus*, Pacific) display many of the characteristics of the phylum. Although the larvae are bilaterally symmetrical, the adults display pentameric radial (five-part) symmetry, a secondarily derived character. Some species display 6- and 12-arm patterns.

PROCEDURE

External Features

1. Observe the aboral (upper) surface of the seastar with a dissecting microscope (Fig. 3.5a). Echinoderms have a calcareous internal skeleton composed of small plates (**ossicles**). Some of the ossicles have short, blunt spines and protrude through the skin. The most conspicuous feature on the aboral surface is the **madreporite,** a round, stonelike plate associated with the water vascular system. The aboral surface of the body is ciliated and covered with short calcareous spines. Around the bases of the spines are fingerlike **dermal branchiae** (respiratory and excretory organs). Minute **pedicellariae** (pincerlike structures) keep the body surface free of debris and small organisms; they should be active in living specimens.

a. Try placing a few drops of particulate suspension near the pedicellariae. Observe and describe their behavior.

2. Turn the specimen so that the oral surface faces you (Fig. 3.5b). The central **mouth** is surrounded by **oral spines** and **podia** (tube feet). Radiating outward from the disc are the **arms.** Running down the center of each arm is the open **ambulacral groove** with its many tube feet (podia) arranged in two to four rows. The podia are the most conspicuous feature of the **water vascular system,** which is a distinctive feature of the phylum. The system consists of a series of fluid-filled tubes that form the basis for the hydraulic system of locomotion. A single modified tube foot—the **tentacle**—occurs at the tip of each ray. It lacks a sucker, and has a light-sensitive pigmented **eyespot** at its base. Try shining a bright light on the eyespot.

b. Does the seastar respond? Describe the behavior.

Feeding

1. Place a living clam in the same dish or aquarium with a seastar that has not been fed for a day or two. Observe the behavior and locomotion of the seastar.

c. How long did it take for the seastar to detect the food and move in its direction?

2. Allow the seastar to feed for a time and then interrupt the activity by gently removing the clam.

d. Is there evidence that the stomach has been everted?

Locomotion

1. Gently stroke the tube feet with your finger or a blunt probe and observe the reaction.

e. Describe the response.

2. Place the specimen upside down in a deep container of seawater.

f. How does the seastar right itself? Which arms (how many) are used? How long does it take? Repeat the observation. Did the seastar use the same arms the second time?

Larval Stages and Development

Fertilization of the seastar takes place externally. The planktonic zygote develops into a bilaterally symmetrical bipinnaria larva (see Fig. 8.1 in Unit 8, and Key D, 10c, in the Keys to Plankton). The bipinnaria larva transforms into a **brachiolaria** larva that, in turn, transforms into a young seastar. Examine slides of seastar larval stages.

Interpretation

1. Are echinoderms bilaterally or radially symmetrical animals? Explain.

2. Oyster harvesters used to drag starfish "mops" over oyster beds to capture seastars. Why did they use this method?

eye spot

madreporite

disk

arms

10 mm

Figure 3.5 Seastar. (a) Aboral view of larger ten-armed star; oral view of smaller star. (b) Oral view. Mouth is ringed by spines and podia (tube feet).

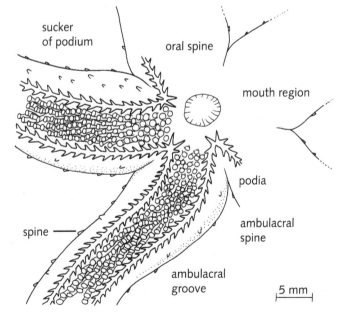

sucker of podium

oral spine

mouth region

podia

ambulacral spine

spine

ambulacral groove

5 mm

3. Oyster harvesters typically chopped up the stars and washed them overboard. Was that a good idea? Explain.

4. Seastars are relatively slow-moving invertebrates, yet many species have broad geographic distributions. How is this possible?

EXERCISE 3

Grazer: Snail

SNAIL
  PHYLUM MOLLUSCA
    CLASS GASTROPODA

Gastropods are motile epifauna. Many, such as limpets and periwinkles, graze or browse continuously on plant or algal material. Others, such as drills, whelks, and conchs, are predators, and use the radular apparatus to drill holes in the shells of bivalve mollusks.

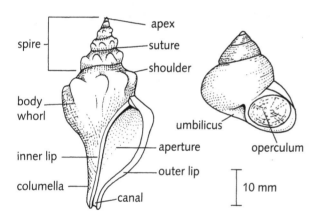

Figure 3.6 External features of gastropod shells with dextral configuration. The apex is pointing up, and the aperture is on the right side and facing you. The specimen on the left has a siphonal canal; the one on the right has an umbilicus, and an operculum sealing the aperture.

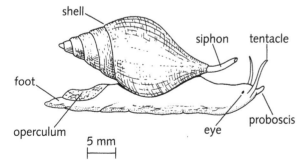

Figure 3.7 External features of an active marine gastropod with a dextrally coiled shell. Note position of operculum on the foot.

The vast diversity in the shells of gastropods is one reason shell collecting is so popular. In this exercise, we will not be able to describe such diversity completely but you will learn some basic terminology of the gastropod shell and examine the unique radula.

PROCEDURE

Shell

1. Examine the shell of your specimen.

   a. Determine if it is **dextral** or **sinistral** in its coiling (Fig. 3.6). Dextral shells coil clockwise (when viewed from the **apex**—the peak of the whorl). Another way to determine direction of coiling is to place the shell on the desk with the **aperture** facing you and the apex pointing away from you. If the aperture appears on the right side, the specimen is dextral; if the aperture is on the left, it is sinistral.

   b. Identify any of the following on your specimen: **umbilicus, columella, canal, spines,** and **sutures.**

   c. Describe the colors and color patterns of the specimen. Can you explain the adaptational value of the structural and color patterns?

Locomotion

1. Observe locomotion in a living snail from the side and from the bottom (observe through the side of the aquarium).

   d. How fast does the snail move?

   e. How does the **foot** appear to move the animal? Do you see any waves of contraction sweeping across the surface of the foot? Which direction are the waves moving relative to the snail's forward motion? Is mucus involved?

2. Note the presence and actions of the **cephalic tentacles,** inhalant **siphon,** extensible **proboscis,** and **operculum** (Fig. 3.7).

   f. Is the shell carried in a straight line with respect to the direction of travel? What part of the shell faces forward?

   g. Gently probe your specimen so that it withdraws into its shell, and then position the shell so that the aperture faces up. Allow sufficient time for the animal to resume normal activity.

h. What is the function of the operculum?

i. Describe how the animal rights itself.

Feeding

1. Observe the specimen as it scrapes food from the surface of the aquarium. This action involves the **radula** (Fig. 3.8), a ribbon of chitinous teeth that moves back and forth over a supporting rod, the **odontophore.** When the animal feeds, this apparatus projects beyond the mouth. The ribbon of teeth moves forward over the leading edge of the odontophore. As the ribbon passes back over the edge, the teeth stand upright, rasp particles from the substrate, and bring the particles into the mouth.

2. Examine prepared slides of radulas, or prepare one from an anesthetized or frozen specimen, as follows:

   j. Amputate the head and remove the muscular buccal mass with scissors and fine-pointed forceps.

   k. Place the buccal mass, containing the proboscis, on a microscope slide. Add a few drops of household Clorox.

CAUTION

Household bleach is sodium hypochlorite, an irritant. Immediately clean up any spills, and wash your hands after you use it.

   When the soft tissue of the proboscis softens, tease the muscles apart with dissecting needles. Within a minute or two, the radula and odontophore will be seen as a J-shaped colorless ribbon.

   l. Rinse the radula carefully in fresh Clorox solution. Use a medicine dropper or Pasteur pipet.

   m. Flatten the radula, and rinse it with one or two changes of 70 percent ethanol. The slide may be made permanent by adding a drop or two of mounting medium and a coverslip.

   n. Examine the radula under the compound microscope.

   o. Sketch the structure. Note the number of rows of teeth, and the direction in which the teeth point.

   p. If radulas from different species are available, compare them. Teeth of grazers should be relatively short and uniform in shape and size. Teeth on the radula of predators are more varied, longer, and sharper.

Larval Stages and Development

In gastropods, fertilization takes place internally. Some species produce ciliated **trochophore** larvae (see Fig. 8.1 in Unit 8, and Key 10, 3a and 8b, in the Keys to Plankton) that develop into the **veliger** stage. The veliger eventually settles out of the plankton and takes up a benthic existence. Other gastropods extrude fertilized eggs into egg capsules. Larval development is bypassed in the capsules; young snails emerge fully formed with a shell. Examine egg capsules on display.

Interpretation

1. Snails are shelled invertebrates. How do they grow, and how does their shell size increase?

2. Essentially all animals fall prey to larger predators. How would the specimen you observed be preyed upon by each of the following:

   a. Crab

   b. Large fish, such as stingray or skate

   c. Seastar

   d. Drill (gastropod)

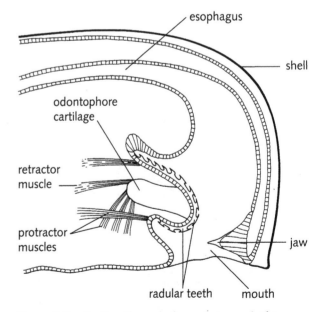

Figure 3.8 Section through the anterior end of a gastropod mollusk, showing the retracted radular apparatus of a gastropod mollusk. During feeding, the apparatus is thrust out through the mouth, and the radular belt glides back and forth over the supporting odontophore.

EXERCISE 4

Grazer: Sea Urchin

SEA URCHIN
  PHYLUM ECHINODERMATA
    CLASS ECHINOIDEA

Echinoids are globose (urchins) or flattened (dollars) echinoderms with radiating spines. Epifaunal and infaunal types live in all ocean basins. Some echinoids are omnivorous, some extensively herbivorous, and others—notably the sand dollars—feed on minute organisms and particulate matter (suspension feeders). This exercise features the herbivorous sea urchins. The more common species include the green *(Strongylocentrotus droehbachiensis)* and purple *(Arbacia punctulata)* urchins of the North Atlantic region, the short-spined urchin *(Lytechinus variegatus)* of the Southeast Atlantic and Gulf of Mexico, and the purple *(S. purpuratus)* and Panamanian *(L. pictus)* urchins of the Pacific coast.

PROCEDURE

External Features

1. Obtain a living specimen and examine the shape and external features (Fig. 3.9a). Is it spherical in shape, or somewhat flattened? Although it may be difficult to see, the external surface of the urchin is ciliated.

2. Examine the movable **spines** that are attached to the plates of the **test** (shell) by a ball and socket joint (Fig. 3.9b). Grasp the tip of a spine.

    a. Do you feel any movement? Observe and describe the response of the spines when you touch the surface of the test with the tip of a probe.

3. **Pedicellariae** (Fig. 3.9c) are varied in form and distributed over the body surface. Study them under the dissecting microscope.

    b. Do the pedicellariae respond to tactile and chemical stimuli? What are their functions? Are they similar in size and shape to those you observed on the seastar?

4. Gaseous exchange in urchins occurs in the fleshy **gills** around the mouth, and the tube feet.

Locomotion

1. Locate the **tube feet.** Note that they are located in five regions **(ambulacral areas)** of the test (Fig. 3.9a).

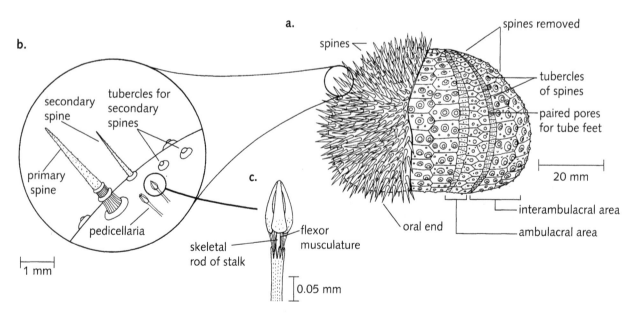

Figure 3.9 External features of a sea urchin. (a) Lateral view, spines removed from half of the specimen revealing pattern of fused plates, and location of ambulacral (location of tube feet) and interambulacral regions. (b) Enlarged portion of test showing spines, spine tubercles, and pedicellaria. (c) Enlargement of one type of pedicellaria.

2. Place your specimen on a glass plate or observe one on the aquarium wall. Note how the spines and tube feet are used in locomotion. Turn your animal upside down and observe how it rights itself.

   c. Which are used more for locomotion, the spines or the tube feet? Describe the righting motion when you turn an urchin upside down.

Feeding

1. Turn your animal upside down and study the mouth area. You should be able to see the tips of five **teeth** that comprise the chewing apparatus of **Aristotle's lantern.** If you examine a demonstration preparation of the apparatus, you will notice the likeness to a lantern. This complex assortment of vertically arranged teeth is supported by an internal framework of skeletal rods, ligaments, and muscles. No other class of echinoderms possesses this mechanism. The mouth is surrounded by a soft membranous area that has specialized tube feet that are probably chemoreceptive.

2. Gently probe the mouth area to provoke movement of the teeth. Notice how they collectively open and close. This action is very effective in scraping plant tissue or algal film from hard substrates. In rock-boring species, the teeth are used to excavate depressions in limestone rock in which the urchins reside.

Larval Stages and Development

Urchins are dioecious; early larval development is similar to that of seastars. However, the terminal larval form of the echinoids is the **echinopluteus** (see Fig. 8.1 in Unit 8, and Key D, 11b, in the Keys to Plankton). Examine a slide of the pluteus. Compare it with the larval stages of the seastar.

Interpretation

1. Urchins tend to have diurnal or nocturnal activity rhythms. Can you document any evidence of these patterns from your readings?

2. Except for a few nations that fish and import sea urchins for human consumption, these animals appear to have relatively little economic value. Sea urchins can create substantial financial impact in certain situations, however. Can you describe any such situation? Do not confine your thinking to direct sale of urchins.

3. What eats urchins? How?

EXERCISE 5
Deposit and Suspension Feeder: Sea Cucumber

SEA CUCUMBER
   PHYLUM ECHINODERMATA
      CLASS HOLOTHUROIDEA

Holothuroideans (= holothurians) have a highly modified echinoderm body plan (Fig. 3.10). They are soft-bodied, bilaterally symmetrical, wormlike animals with recognizable anterior and posterior ends.

PROCEDURE

Locomotion

Some sea cucumbers have the essential components of a hydrostatic skeleton that enables them to burrow in the substrate. Tube feet of burrowing forms are reduced in size or number, or are absent. Locomotion through the substrate is wormlike. Like burrowing polychaete worms, peristaltic waves of contraction pass along the body wall.

1. The **tube feet** used for locomotion and attachment are confined to five **ambulacral bands** running the length of the body.

   a. Are all of the tube feet identical in size? If you observe that certain ambulacral bands have larger podia than others, can you relate that to function?

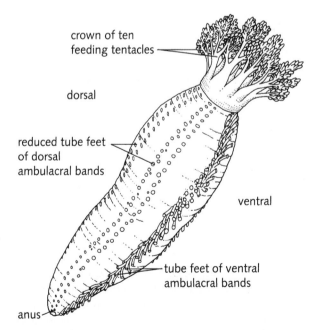

crown of ten feeding tentacles

dorsal

reduced tube feet of dorsal ambulacral bands

ventral

tube feet of ventral ambulacral bands

anus

Figure 3.10 Epibenthic sea cucumber. External morphology.

Feeding

If you have a living specimen that is active, observe the feeding sequence. Be patient—sea cucumbers are slow and cumbersome, and tedious to watch.

The oral tube feet (Fig. 3.10) are highly modified **feeding tentacles** that can be withdrawn inside the mouth. They are coated with sticky mucus that picks up food particles. A few tentacles grope for particles in the substrate, while others are inserted in the mouth. Any organic material on the particles is digested as the sediment passes through the gut. In essence, sea cucumbers function somewhat like huge earthworms.

Internal Anatomy

There is no need to dissect a sea cucumber, but you should be aware of the unique gaseous exchange organs, the **respiratory trees.** These paired, highly branched, muscular tubes are connected to the **cloaca,** a chamber just inside the anus. Water is pumped into and out of the system through the anus and cloaca. Gas exchange occurs through the walls of the tubular network.

Larval Stages and Development

Holothuroidean larval development resembles that of the other echinoderms, but, as in the other groups, the larval stages (**pentacula** and **auricularia**) are unique (see Fig. 8.1 in Unit 8, and Key D, 10b,c, in the Keys to Plankton). Specimens are rarely available for examination.

Interpretation

Describe the contribution that sea cucumbers make to the food web. How do they fill an important "link"? What is the origin of their food? Do they have any known predators?

E X E R C I S E   6
Deposit and Suspension Feeder: Clam

CLAM
  PHYLUM MOLLUSCA
    CLASS BIVALVIA

Clams, oysters, mussels, and scallops (bivalves) are shelled, sedentary, suspension-feeding invertebrates. Most clamlike mollusks are infauna in that they live beneath the surface of the substrate. Oysters and mussels, however, are epibenthic because they attach to the surface of the substrate. Scallops are also epibenthic, although they only lie upon the substrate—they are not anchored to it. In fact, they are quite mobile.

All of the bivalves possess **ctenidia,** enormously enlarged, ciliated gills that also function as the major organ of food collection. The ctenidia are much larger than is required solely for the respiratory needs of the animal.

PROCEDURE

External Features

1. Select a specimen and examine the external features of the shell (Fig. 3.11).

    The typical bivalve mollusk (oysters are atypical) has a symmetrically clam-shaped shell with a springy, proteinaceous **hinge** at the dorsal edge, between the **umbos,** which are the oldest parts of the **valves.** In the living mollusk, the hinge is resilient and forces the valves apart when the adductor muscles are relaxed. The shell is secreted and maintained by the **mantle** (discussed later in this exercise). The outer surface of the valve may be covered with a **periostracum,** which is continuous with the outer mantle edge. It often is worn away, particularly in specimens living in very abrasive substrates.

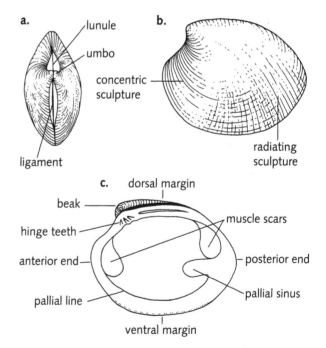

Figure 3.11 Features of bivalve shells. External features from dorsal (a) and lateral (b) views. (c) Internal aspect showing teeth and "scars."

a. Note the presence of any external ridges, spines, or folds.

b. Is there evidence of predatory attack (for example, drill holes or chips along the ventral margin of the shell)? Note any coloration.

Internal Anatomy

1. Orient your specimen to correspond to the position illustrated in Figure 3.12. Open the clam with a clam or oyster knife (have the instructor demonstrate this for you). Carefully insert a scalpel (or the clam knife) inside the shell at the posterior margin and cut the posterior adductor muscle (Fig. 3.12). Repeat the operation at the anterior end.

2. Lift the left valve (the one facing you) and carefully separate the layer of **mantle** tissue from the shell. Allow the mantle to resume its normal position—that is, covering the soft body of the organism lying in the left (lower) shell. Identify the exposed **adductor muscles.**

3. Carefully cut away the left mantle lobe to expose the mantle cavity and the organs it contains. Locate the following structures: **ctenidia** (gills), **labial palps,** and both dorsal, **excurrent** and ventral, **incurrent siphons.**

    c. How long is the siphon in your specimen? Can you relate siphon length to the depth in the substrate at which it normally resides?

4. Examine the arrangement of the ctenidia and their relation to the **foot** and to the mantle. Note that each ctenidium (right and left sides of the mantle cavity) consists of two parts (inner and outer **demibranchs**) (Fig. 3.12).

5. If your specimen is living, place a few drops of particle suspension (congo red–yeast preparation) on the surface of the ctenidium.

    d. Do the particles move? In which direction? Is mucus involved in the mechanism?

    e. Food material is moved toward the labial palps and then into the mouth; discarded particles are moved off the ctenidia in mucus strings. Do your observations match this description?

6. Remove a 5-mm square of tissue from the edge of the gill and make a wet mount (with seawater) on a microscope slide. Examine it under low magni-

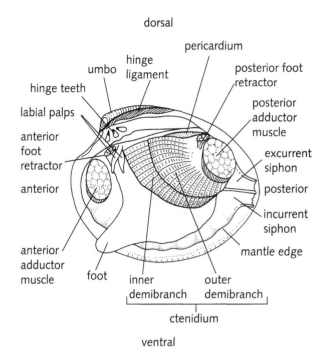

Figure 3.12 Hard clam with left valve and mantle removed to expose viscera.

fication. Locate the slitlike openings through which water flows. Switch to high-dry magnification and note the location, arrangement, and beating of **cilia,** the driving force of the water current passing over and through the gills. Ciliary action also moves material to food grooves located on the ventral margin of each ctenidium.

Locomotion

1. Place a fresh specimen on a sandy substrate in the aquarium.

    f. Describe how the animal uses its muscular **foot** to burrow into the substrate. How does it right itself? How long does it take? Comment on the animal's vulnerability during this maneuver.

Exceptions to the Rule

Certain bivalves exhibit modifications of the above body plan. In the blue mussel, *Mytilus* sp., the foot is reduced in size and contains a **byssal gland.** The gland secretes filaments that secure the animal to a hard substrate (Fig. 3.13a). Sizes of the anterior and posterior adductor muscles are noticeably different.

The oyster, *Crassostrea* sp., lacks a muscular foot entirely and has only one adductor muscle (Fig. 3.13b). The oyster is cemented to a hard substrate or to other oysters when the larval stage settles out of the plankton.

Larval Stages and Development

In bivalves fertilization takes place externally. The **trochophore** larval stage is barrel-shaped and contains ciliated bands (see Fig. 8.1 in Unit 8, and Key D, 3a, in the Keys to Plankton). The more advanced, uniquely molluscan **veliger** stage is more complex. Examine prepared slides of these stages.

Interpretation

1. Clams are shelled invertebrates. How do they grow?

2. Identify three predators of clams. Briefly describe how the predators devour the clams.

3. What structure is unique to bivalves?

4. Which are more vulnerable to predators: edible mussels, sea scallops, or clams? Weigh the structural and behavioral advantages and disadvantages of each mollusk, and consider their habitats as well. Defend your answer.

5. What is mucociliary feeding? Do all suspension feeders obtain food by a mucociliary mechanism?

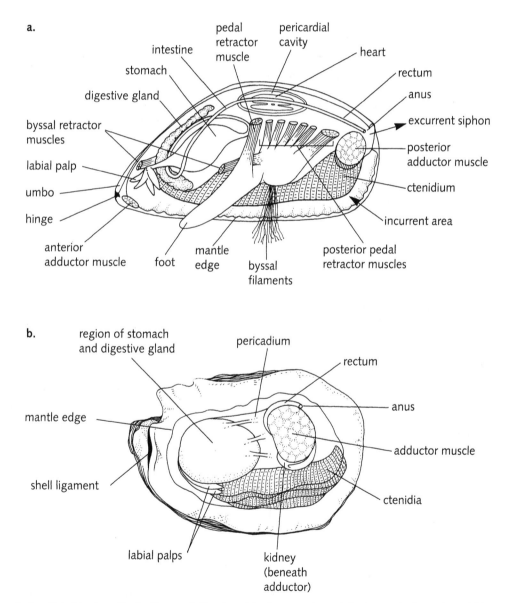

Figure 3.13 Epibenthic bivalves illustrating variation in adductor muscle and foot morphology. (a) Edible mussel with unequal adductor muscles, and reduced foot with byssal gland. (b) Oyster with one adductor muscle (monomyarian condition), and no foot. Animal shown in bottom shell.

Benthic Meiofauna

OBJECTIVES

After completing this unit, you will be able to

- Perform techniques for studying live microscopic organisms;

- Become familiar with the morphology, locomotion, and behavior of some meiofaunal representatives;

- Appreciate the size relationships of meiofauna with that of sediment grains and the interstitial spaces in which the meiofauna live; and

- Detect and study meiofauna in natural sediment samples.

INTRODUCTION

The distribution of benthic species ("benthic" means "to live in or on the sea bottom") is controlled principally by the composition of the substratum (described below). Moreover, morphological, feeding, and physiological adaptations of benthic species are related to the substratum makeup.

Hard substrata present a surface to which organisms attach, or into which they bore. These substrata consist of a single material, such as rock, hard skeletons (corals), wood, or recemented sedimentary grains (lithified coral rubble). Species inhabiting such substrata are discussed in Unit 11 on fouling organisms.

Soft substratum consists of: (1) sedimentary particles of varying size, form, and origin (mineral grains, remains of animal shells and skeletons); (2) organic particles of varying size, composition, and origin (dead and decaying plants and animals); and (3) water-filled spaces among the particles (**interstitial spaces**).

The kinds of benthic organisms in residence correlate with the type of soft substratum. For example, in a sediment consisting of sand particles 2 mm in diameter, larger burrowing **macrofauna** (larger than 1.0 mm) push through sedimentary grains whose diameters are small relative to the body size of the burrower. Benthic **meiofauna** (lesser animals)—those organisms that pass undamaged through 0.5-mm to 1.0-mm screens, but are trapped on 42-μm to 100-μm screens—can travel through the narrow interstitial spaces among sediment grains larger than themselves. Note that although the term "interstitial fauna" is often used synonymously for meiofauna, the former term is acceptable only if the fauna move through the interstitial spaces with minimum disturbance of the particles. Many silt- and mud-dwelling animals would not really be classified as interstitial because they cannot navigate through the extremely small water-filled spaces.

Sediment composition and texture control the type and abundance of meiofauna. Two taxonomic groups typically dominate meiofaunal communities: nematodes (see Fig. 4.8) constitute more than 50% of the total meiofauna; harpacticoid copepods (Fig. 4.1) are second in abundance. These taxa are then followed by a bewildering array of species.

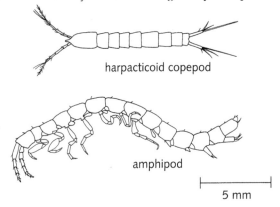

harpacticoid copepod

amphipod

5 mm

Figure 4.1 The harpacticoid copepod and the amphipod shown here possess the vermiform shape typical of meiofauna. They differ markedly from benthic and planktonic forms (see Fig. 8.3 in Unit 8, and Key D, 6c, in the Keys to Plankton).

The majority of meiofauna exist in the upper 2 cm of sediment, but in some sandy beaches organisms can live at depths greater than 10 cm. In muds and heavy organic detrital sediments, the organisms are confined to the upper few millimeters.

The interstitial habitat has selected for an elongated and wormlike (**vermiform**) body shape in some 37 different taxa of meiofauna. For example, the body shapes of interstitial amphipods and harpacticoid copepods (Fig. 4.1) differ from those of benthic or planktonic forms (see Fig. 8.3 in Unit 8, and Key D, 6c, in the Keys to Plankton).

Meiofauna can occur in large population densities: populations of 10^5 to 10^7 individuals per square meter are typical in sediments on the continental shelves. Such abundant populations are supported by microscopic organisms that thrive in the interstices and on the surface of sedimentary grains. Meiofauna, in turn, are preyed upon by a variety of sedimentary fauna (Fig. 4.2).

This unit is divided into four parts. The first part introduces basic techniques for the examination of living microscopic organisms. The second part describes locomotion and morphology of representatives of several common meiofaunal taxa. The third part relates the size of these meiofauna to that of the sediment grains and the interstitial spaces in which they live. The fourth part describes some of the techniques used to analyze sediments containing meiofauna.

Figure 4.2 Trophic relationships of microbiota, meiofauna, and macrofauna of sedimentary substrates. Food input to the interstitial system includes: (1) primary production by microflora (for example, diatoms), and (2) dissolved and particulate organic matter that is flushed through the sediment. Primary users of these organics are bacteria in the water and attached to the grains (a). Other microorganisms, including diatoms, blue-green algae, and protozoans, inhabit the irregular surface of intertidal sand grains, and provide a source of nutrition for meiofauna. These microorganisms, along with plankton settling on the sediments, form the basis for food webs that involve meiofauna (b), macrobenthic infauna (c), and epifauna (d). Feeding mechanisms of the grazers and predators are also indicated.

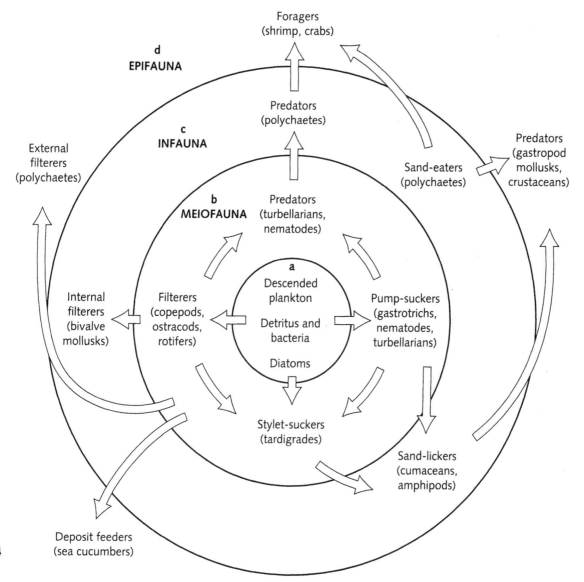

EXERCISE 1

General Laboratory Techniques

In this unit, you will deal with living cultures of organisms that require special handling for microscopic observation. Several techniques for handling these cultures are described below. Take every precaution to *avoid contaminating living cultures*. Each culture will have its own pipet for dispensing. Use no other pipet!

PROCEDURE

1. Obtain compound and dissection microscopes, a lamp, and an assortment of the following items: standard microscope slides and coverslips; depression slides; a medicine dropper or Pasteur pipet; a teasing needle; a fine soft (camel's hair) brush, bamboo splinter, or sharpened toothpick; and a small petri dish or Syracuse watch glass.

2. Calibrate your microscope so that you can measure microscopic specimens and substrate particles. Refer to Appendix I for further details.

3. Select a slide preparation as directed in the description for each specimen.

   Wet mount (Fig. 4.3a): for routine *short-term* study of fresh, living material that is thin or microscopic. These preparations dry out rapidly as the water evaporates.

   a. Use a slide, coverslip, and teasing needle.

   b. Place a drop of the culture on your slide.

   c. Touch the coverslip to one edge of the drop, and gently lower it with the teasing needle. If you are careful, you will not get air bubbles under the coverslip. Air bubbles are seen as dark-edged circles under magnification.

   Vaseline (petroleum jelly) mount (Fig. 4.3b): for routine *long-term* study of fragile or relatively large/thick specimens.

   a. Use a slide, coverslip, Vaseline, and a toothpick.

   b. Apply four narrow (1–2 mm thick) Vaseline walls in the shape of a square on the slide with the toothpick. Make the square reservoir the same size as your coverslip. The walls raise the coverslip to prevent crushing large specimens, and they seal the preparation to prevent evaporation.

   c. Fill the reservoir with culture (2–3 drops).

   d. Place the coverslip on the Vaseline walls.

   Depression slide (Fig. 4.3c): for *long-term* study of relatively large and highly motile specimens.

   a. Fill depression in the slide with the culture.

   b. Application of a coverslip is optional. Some preparations are observed better without a cover. Consult your instructor for advice. *NOTE:* Most depression slides are thicker than conventional slides, which reduces the working distance of the microscope. Extreme caution must be taken when changing from low-power to high-power objective.

   Squash preparation (Fig. 4.3d): used to separate cells, or to spread component parts of specimens to better observe structure.

   a. Use a standard slide, coverslip, teasing needle, and paper towel.

   b. Prepare a wet mount (Fig. 4.3a) as described above.

   c. Place the paper towel on the laboratory bench.

   d. Invert your slide on the paper towel (Fig. 4.3d) and push downward with your thumb over the coverslip.

   e. Examine your slide with the microscope. If the preparation did not spread adequately, repeat step (d), but rotate the slide slightly as you apply thumb pressure.

   Slow-Motion wet mounts: for retarding motion of rapidly moving organisms to allow detailed observation of behavior, locomotion, and structure. Three alternative techniques are described here.

   a. Apply a drop or two of a solution of **methyl cellulose** or an equivalent product *(Protoslo, Detain)* to a wet mount (Fig. 4.3a) or depression slide preparation (Fig. 4.3c) and add (or replace) the coverslip. The organisms are slowed by the increase in the viscosity of the medium.

   b. Apply a ring of **methyl cellulose** or equivalent to a slide. Place a drop or two of the culture on the center, and add a coverslip. The slowing agent will gradually diffuse into the preparation.

c. Prepare a wet mount (Fig. 4.3a), but use a small piece of Saran Wrap (Dow Chemical Company, Midland, Michigan) instead of a glass coverslip. The **plastic wrap** holds the organism relatively still and retards evaporation, but allows gas exchange to occur. Objectives up to and including oil immer-sion can be used with this material. Specimens may also be mounted between two pieces of plastic wrap and then placed on a slide. Such preparations are easily flipped over to observe the opposite side of the organism.

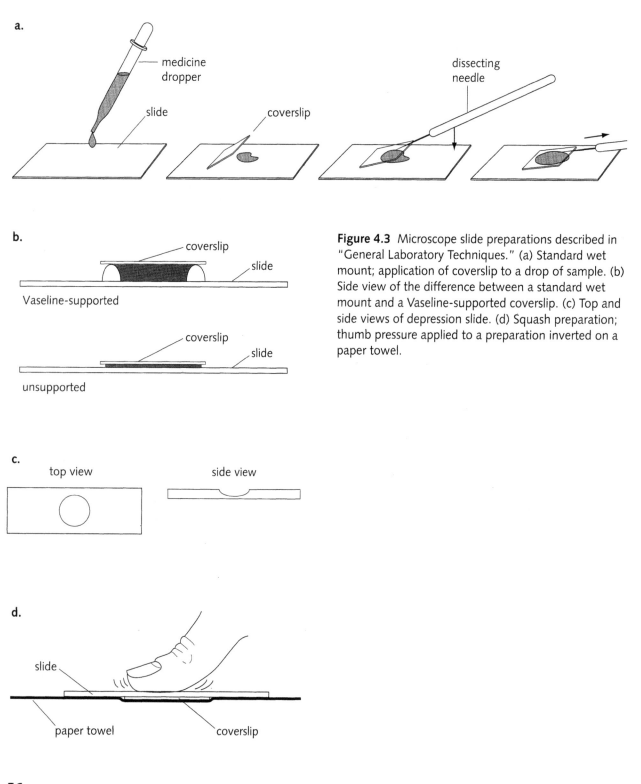

Figure 4.3 Microscope slide preparations described in "General Laboratory Techniques." (a) Standard wet mount; application of coverslip to a drop of sample. (b) Side view of the difference between a standard wet mount and a Vaseline-supported coverslip. (c) Top and side views of depression slide. (d) Squash preparation; thumb pressure applied to a preparation inverted on a paper towel.

A. Representatives of Meiofauna

Seven species of meiofauna are described in this unit. One other representative (copepod crustacean) is described in the plankton unit (Unit 8).

You will be advised which species are available for study, and the extent to which you will examine each.

GENERAL PROCEDURE

Follow the procedures and descriptions specific to each specimen and refer to the illustrations provided. Record your observations and make drawings in your notebook, or on separate pieces of drawing paper, as directed by your instructor. Be sure to answer the questions that you encounter in the descriptions.

Use the following overview (**checklist**) of the tasks common to all specimens.

1. Locomotion and movement

    a. First observe living specimens at normal speed in routine wet mounts, Vaseline mounts, or depression slides. If you then need to retard movement, use the slow-motion applications.

    b. Sketch and describe the changes in the specimen's shape as it moves. Make a series of outline drawings that show the sequence of shape changes and the progress of movement. Use arrows to indicate any forward progress or direction of movement. Be sure to include some indication of size (direct measurement or scale bars) in your drawings.

    c. Slow the animal, if necessary, and observe features not readily seen at normal speed.

2. Morphology and size

    a. Observe specimens (stained or unstained) in slow-motion mounts as recommended for each species.

    b. Draw, as accurately as possible, the conspicuous external and internal features of each species. Indicate the magnification at which you made the observations and whether the specimen was living or preserved, stained or unstained.

    c. Measure the **length** and **width** of a specimen of each species, and record the measurements on your drawings.

3. Feeding and other behavior.

    Describe any observed feeding activities, or other behavior of interest to you. Make drawings as necessary.

EXERCISE 1
Phylum Ciliophora

Individuals of this phylum of unicellular organisms are extremely common in aquatic habitats. Ciliophorans (= ciliates = hair-bearing) are characterized by a relatively rigid, or fixed, cell shape with external **cilia** for locomotion. Cilia associated with a distinct **cytostome** (= cell mouth) also produce feeding currents. Each cell body holds two nuclei: a small **micronucleus** for reproduction, and a larger **macronucleus** that controls normal cell functions.

PROCEDURE

1. Prepare a wet mount (Fig. 4.3a) of *Spirostomum* sp. (Fig. 4.4) culture. Observe (and sketch) movements of this animal under low and high magnifications. Note that one end is blunter than the

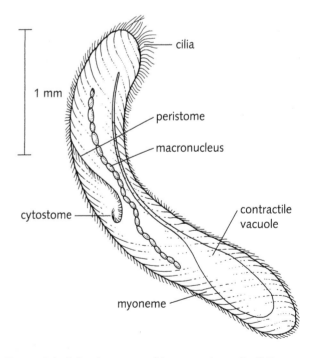

Figure 4.4 *Spirostomum ambiguum*, a large (to 3.5 mm) unicellular ciliated organism (phylum Ciliophora).

other. Which end usually moves forward? Is this a fast-moving animal?

2. Prepare a slow-motion mount, if necessary. Measure the **length** and **width** of a specimen. Some *Spirostomum* sp. organisms are known to reach a length of 3.5 mm. How does the length of your specimen compare with that measurement?

3. Observe beating of the **cilia.** Do they cover the entire body? Try to identify the **cytostome** and **peristome,** the organelles through which food is directed into the cell (Fig. 4.4).

4. *Spirostomum* sp. contracts its body with contractile bands **(myonemes)** (= muscle threads) that have a spiral pattern (Fig. 4.4). Can you observe changes in body shape?

5. You may see the **macronucleus** and the expansion and contraction of the very large **contractile vacuole** in the cell. This organelle, near the posterior end of the cell, is responsible for water regulation.

Interpretation

1. Perhaps your culture of *Spirostomum* sp. was not a pure culture, in that other smaller unicellular organisms were present. Could these organisms possibly serve as food for *Spirostomum* sp.? What was the relative difference in size between the two organisms?

2. What is the ratio of cell length to width? Does this ratio agree with the "wormlike" designation of interstitial fauna?

EXERCISE 2
Phylum Platyhelminthes

This phylum includes three classes of worms, two of which are entirely parasitic. Members of the class Turbellaria are free-living. They are dorso-ventrally flattened and **acoelomate** (lacking a body cavity; a loose tissue fills the space between the internal organs and body wall). The **mouth** is the only opening of the **digestive tract** (when one occurs). **Protonephridia** (= first kidneys) are present, and the reproductive system is **monoecious** (= one house; ovary and testis in the same individual).

Flatworms vary in shape from oval to elongate. Head projections may occur in some species. Species range in size from microscopic forms to some that are more than 60 cm long. Most are less than 10 mm in length, however.

Turbellarians are primarily aquatic, and the majority are marine. Most of these are benthic, living in substrates, on shells, or on seaweeds. Many species commonly inhabit interstitial spaces.

PROCEDURE

1. Examine living specimens of the microscopic *Stenostomum* sp. (Fig. 4.5) in a depression slide (Fig. 4.3c) or a Vaseline mount (Fig. 4.3b). Observe (and sketch) movements of this freshwater animal. Can you observe beating of **external cilia?**

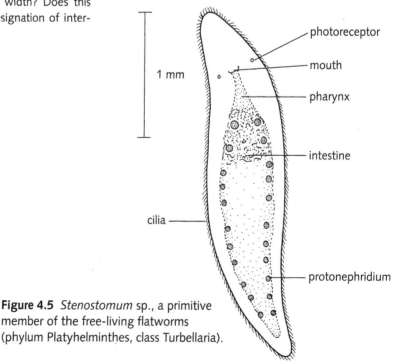

Figure 4.5 *Stenostomum* sp., a primitive member of the free-living flatworms (phylum Platyhelminthes, class Turbellaria).

How would you estimate the relative speed of travel of this animal? Would you rate it faster or slower than *Spirostomum* sp.?

2. Prepare a slow-motion mount, if necessary, and attempt to identify the simple **pharynx**, ciliated saclike **intestine, protonephridia,** and external **cilia** (Fig. 4.5). Draw and measure the **length** and **width** of the specimen.

Interpretation

1. Compare *Spirostomum* sp. with *Stenostomum* sp. in terms of size, shape, locomotion (speed), and observed morphology. If you were not told that they belong to different phyla, how could you distinguish between them?

EXERCISE 3

Phylum Gastrotricha

Some 450 known marine and freshwater species of gastrotrichs (stomach hair) exist. They are bilaterally symmetrical animals that glide with their ventral surface in contact with other organisms or particles. Three characteristics are obvious (Fig. 4.6): a **forked tail** containing **adhesive glands;** a modified surface with **spine- or scalelike** structures covering the dorsal side; and a patterned distribution of **cilia** that are usually restricted to the ventral side (hence the name of the phylum). Most adults are less than 1 mm in length.

PROCEDURE

1. Examine a culture of *Lepidodermella* sp. (Fig. 4.6) in a depression slide (Fig. 4.3c) or Vaseline mount (Fig.4.3b). View them first with a dissection microscope, and then with a compound microscope. Observe and record their movements. How do you compare this organism's speed with that of *Spirostomum* sp., or *Stenostomum* sp.? Is the forked tail used in any particular way?

2. Prepare a slow-motion mount *without a coverslip* and check for their presence under low power (via a compound microscope). When you have one or more specimens, apply a single drop of Neutral Red stain and add a coverslip. Relocate the specimens and examine them under high-dry magnification. Note the **scaly** appearance.

With appropriate lighting you should see the **anterior sensory bristles, mouth, pharynx, gut, adhesive glands,** and portions of the **reproductive system** (Fig. 4.6).

3. Measure and record the **length** and **width** of a specimen.

Interpretation

1. Did you observe a wide range of sizes for the gastrotrichs in your culture? What might this indicate, given that males and females are equal in size?

2. What is the function of the forked tail?

3. Is this a fast-moving animal? Explain. Compare its speed with that of other species you may have observed.

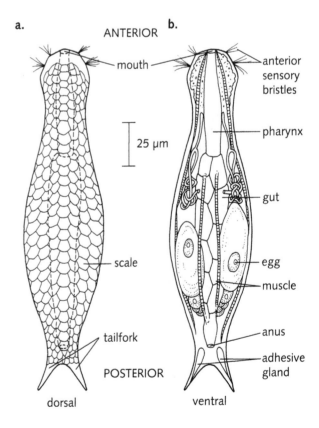

Figure 4.6 Phylum Gastrotricha. (a) Dorsal and ventral views of *Lepidodermella* sp. (b) Internal anatomy of a generalized specimen.

EXERCISE 4

Phylum Rotifera

Rotifers (= wheel-bearer) are bilaterally symmetrical animals with two distinct features: a complex set of **ciliated** organs **(corona)** used for locomotion and feeding; and a muscular **pharynx** equipped with a complex set of jaws (Fig. 4.7). The name "wheel-bearer" refers to the illusion of a rotating wheel created by the metachronal beat of the coronal cilia. Most rotifers are small (less than 1.5 mm) and solitary; they move freely, either by swimming or crawling.

PROCEDURE

1. Examine a culture of *Brachionus* sp., a marine rotifer, under a dissecting microscope. Place a few rotifers on a depression slide (Fig. 4.3c) or Vaseline mount (Fig. 4.3b). Locate the animals at low power. Two markedly different sizes may be present. The larger forms are females, while the decidedly smaller ones are male. Study the larger ones in your sample.

2. Observe and record the locomotory activities of the rotifer. Make a series of outline sketches to illustrate changes in body shape as the animal moves and feeds. Observe how it uses its "feet," which are equipped with adhesive glands. Is this animal fast?

3. Prepare a second depression slide and place a small drop of food suspension (Congo red–yeast mixture) at the edge of the culture. Observe the rotifers under low power. Track any particles that become caught in the coronal currents. Describe the ingestion and movement of particles in the gut.

4. Prepare a slow-motion slide of the culture. Select a large specimen and measure its **length** and **width.**

5. Read the following description and try to locate the described features on your specimens (Fig. 4.7). A rotifer is divided into three regions: **head** (with **corona**), **body**, and **foot**. The corona contains two pairs of ciliated rings used in feeding. Some enlarged cilia are fused into stout **cirri.** Under high power, locate the **pharynx, stomach, body musculature, cement glands,** and the **ovary** and **eggs** (in females).

Interpretation

1. How does a rotifer use its "feet" and "toes"? Do they serve a function similar to the use of the forked tail of the gastrotrich?

2. Compare the use of cilia in the rotifer with that of other species that you observe. Consider the presence or absence of cilia, and the functions.

3. Rank the speed of travel of a rotifer relative to that of other species studied.

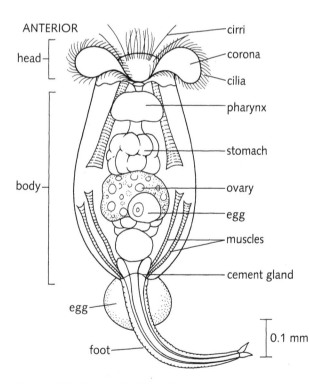

Figure 4.7 Phylum Rotifera. Female specimen of a marine species, *Brachionus* sp.

EXERCISE 5

Phylum Nematoda

Threadlike round worms include 15,000 species. They are usually the most abundant metazoans in marine sediments. Typically they are cylindrical and taper at both ends (Fig. 4.8), and range from less than 1 mm to several meters in length. A resistant **cuticle** covers each organism—a feature unique to the group.

PROCEDURE

1. Obtain some specimens from a culture of vinegar worms, *Turbatrix aceti,* and place the sample in a depression slide (Fig. 4.3c) *without a coverslip.* Note the distinctive sinuous and thrashing movements. Does this animal make any significant forward progress? If so, does it move forward rapidly or slowly?

2. Add a drop of slowing agent to the slide, and repeat your observations. Describe the movement, and make a series of sketches to illustrate it. Measure and record the **length** and **width** of a specimen.

3. Examine a prepared slide of a nematode (Fig. 4.8) under low and high power on a compound microscope, and locate the **mouth, muscular pharynx,** and **intestine.** The **ovary** and large **eggs** may be visible in the female.

Interpretation

1. The habitat of vinegar worms differs considerably from that of interstitial nematodes. Other than the fact that vinegar worms live in acetic acid rather than seawater, what other differences in habitat exist between vinegar worms and meiofaunal nematodes?

2. Based upon your answer to question 1, and the supposition that you could place vinegar worms in a sedimentary substrate, would you expect them to move in the same way that they do in the unconfined liquid medium?

3. Rank the forward speed of progress of the vinegar worms relative to that of other meiofaunal species studied.

EXERCISE 6

Phylum Annelida

Most marine and estuarine annelids are polychaetes (class Polychaeta = many bristles), and most of those are conspicuous macrobenthic forms. Some annelids, however, qualify as meiofauna. The representative used in this exercise is *Aeolosoma* sp. (Fig. 4.9), a freshwater oligochaete (class Oligochaeta = few bristles) found in mud and debris of stagnant pools.

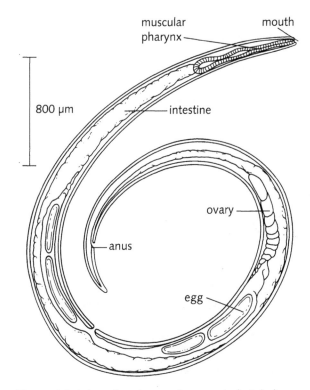

Figure 4.8 Internal anatomy of a nematode (phylum Nematoda). Details are evident only in relatively large and stained specimens.

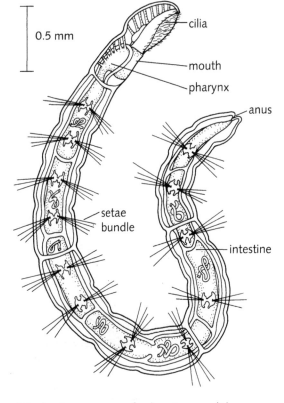

Figure 4.9 *Aeolosoma* sp., a freshwater annelid (phylum Annelida; class Oligochaeta).

1. Examine specimens of *Aeolosoma* sp. on a depression slide (Fig. 4.3c).

2. Describe the locomotion, which usually involves a crawling movement on, or in, the superficial layers of the substrate. Observe the **bundles of setae** on nearly every segment; they secure a hold on substrate particles. Some species swim with the aid of the beating of the cilia in the anterior end of the worm. Is this worm a slow or fast crawler?

3. Slow the worms, if necessary, to measure **length** and **width,** and to observe their morphology (Fig. 4.9). **Metamerism** (body segmentation) is well developed, and **setal bundles** are present (the worms lack the parapodia common to polychaete annelids). The **ciliated** anterior end of the worm does not include sensory appendages. Most aquatic oligochaetes ingest substrate with the **mouth** and **pharynx;** the organic component is digested as it passes through the **intestine.**

Interpretation

1. Nematodes typically possess only longitudinally arranged muscle bands in the body wall. Annelids have at least circular and longitudinal muscle arrangements. Relate these differences in body structure with the different kinds of movements that you observed in these worms.

2. Rank the speed of travel of this species relative to that of other meiofauna studied.

EXERCISE 7
Phylum Tardigrada

About 400 species of tardigrades (slow walk), or "water bears," are known. Most are freshwater forms, but marine species exist in interstitial spaces among sedimentary particles and on the surfaces of aquatic plants. The term "water bear" derives from the bearlike appearance of the short body and the pawing motion of the short, stubby appendages (Fig. 4.10).

PROCEDURE

1. Prepare a depression slide (Fig. 4.3c) or a Vaseline mount (Fig. 4.3b) of the tardigrade culture.

2. Observe the movements of a specimen with a compound microscope. Describe how your specimen uses the appendages in locomotion. Describe any feeding activity, however slow it may be.

3. Measure the **length** and **width** of a specimen and identify the conspicuous external features (Fig. 4.10). Is there any evidence of what appears to be armor plating on the body? Observe the **legs,** each of which has two or more **toes, claws,** or some combination of both. Increase the magnification and draw one of the legs, showing the toe and claw feature of your specimen. Can you see the **claw gland?** Does your specimen have eyespots or any spines protruding from the anterior end?

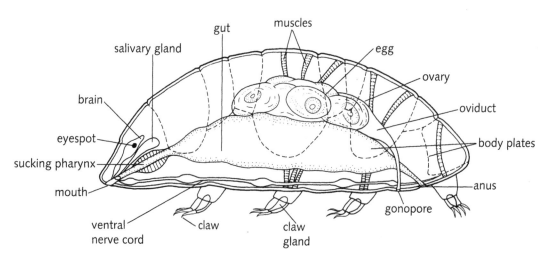

Figure 4.10 Phylum Tardigrada. External and internal features of a water bear, lateral view.

4. Try to identify the internal organs (Fig. 4.10). The muscular **sucking pharynx** is often observed as a dense organ in the anterior end that is flanked by a pair of **salivary glands.** If the animal is actively feeding, you may be able to detect the use of a **stylet** and see parts of the **gut** filled with food. **Eggs** are often conspicuous in the **ovaries** of females. A **ventral nerve cord** and paired **ganglia** run the length of the body. Several **muscle fibers** extend from one part of the body wall to another. If you can locate any of these muscle fibers, trace them from one end to the other. Predict what will result from the contraction of any one fiber. Can you relate this action to any body movements observed earlier?

Interpretation

1. How is locomotion of the tardigrades similar to that of the rotifers and gastrotrichs?

2. Some authors suggest that tardigrades are "arthropodlike" animals. What features could support that statement?

3. Is this the slowest animal you observed in this unit? Explain.

Summary Interpretation (All Specimens Included)

These questions and problems are designed to provide you with a comprehensive assessment of the meiofauna.

1. Construct a table that summarizes features of the specimens that you studied. List the specimens in the first column. Label the following columns: *Phylum* or *Class; Size; Locomotory Mechanism; Relative Speed; Diet.* Fill in the blanks with brief descriptions.

2. Rank the specimens according to size (body length).

3. Determine the volume of each of the specimens for which you measured length and width. Conversion factors (**C**) can be used in the equation

$$V = L \times W^2 \times C$$

where V is body volume in nanoliters (nL), L is length (mm), and W is maximum width (mm).

The following conversion factors (**C**) are from Feller and Warwick (1988):

Hydroids	385	Oligochaetes	530
Turbellarians	550	Tardigrades	614
Gastrotrichs	550	Ostracods	450
Kinorhynchs	295	Tanaids	400
Nematodes	530	Mites	399
Polychaetes	530		

4. Based upon your volume determinations, determine the potential abundance (per square meter of surface) of one of your species in 0.5 m² of substrate, 0.5 cm deep, where interstitial water occupies 30% of the substrate. Compare your calculations with the densities mentioned in the introduction to this unit.

   a. Calculate the population density in terms of number of individuals per mL of interstitial water. There are 1000 nL in one microliter (μL), and 1000 μL in one milliliter (mL), or 1 mL = 10^6 nL.

   b. Determine the volume occupied by 0.5 m² of substrate, 0.5 cm deep, with 30% water composition.

   c. Calculate the number of individuals present in the interstitial water of the square meter sample.

   For example, Species P (a nematode) is 0.75 mm long and 0.1 mm wide. How many exist in an area of 1 m², 1 cm deep, with a water composition of 50%?

   V (of P) = (0.75 mm × 0.1 mm²) × 530 = 3.98 nL

   Number of P/nL = 10^6 nL/3.98 nL = 2.5126 × 10^5 individuals/nL

Volume of substrate + water = 1 m ×
1 m × 0.01 m = 0.01 m^3
0.01 m^3 = 10000 cm^3 = 10000 mL

Volume of water = 0.5 × 0.01 m^3 = 0.005
m^3 = 5000 mL

Number of P/m^2 (to depth of 1 cm) = 5000
mL × (2.5126 × 10^5) = 1.2563 × 10^9/m^2

5. This series of problems involves the concept of surface area:volume ratio.

a. Calculate the surface area (SA) of the ciliophoran *Spirostomum* sp. Assume that the cell is cylindrical in shape. Area (A_b) of the barrel of the cylinder can be calculated with the equation

$$A_b = \pi \times d \times L$$

where π = 3.14, d is the diameter (= body width), and L is body length.

Area of the two ends of the cylinder can be calculated as

$$A_e = 2 \times (\pi \times r^2)$$

where r is ½ the diameter (= body width).

Add the two area calculations together to get total surface area:

$$SA = A_b + A_e$$

b. Calculate the volume (V) of the cylinder:

$$V = \pi \times r^2 \times L.$$

c. Calculate the surface area: volume ratio:

$$SA/V = ?$$

d. Now perform the same calculations using the same body length, but double the width (= d). How does the SA/V ratio change?

6. Indicate which specimens are smaller and larger than the grains of sediment that you observed. Did you observe any specimen that can ingest the grains?

7. Construct a food web of the specimens observed. Identify the primary producers, herbivores, carnivores, and so on.

B. Sediments and Interstitial Spaces

EXERCISE 1

Comparing Size of Meiofauna with Environment

PROCEDURE

1. Place a thin smear of petroleum jelly (Vaseline) on a microscope slide.

2. Obtain a sample of dry marine sediment (silt, quartz sand, carbonate sand, aquarium gravel), and sprinkle a layer of it on the smear. Tap the slide to remove excess material.

3. Measure several grains and the interstitial spaces (use an ocular micrometer).

4. In your notebook or on a sheet of blank paper, draw a square 10 cm on each side. Fill the square with *outline* drawings of contiguous grains of sediments and their associated interstitial spaces. Place a reference scale bar of 0.1 mm or 0.5 mm (use the information from step 3)

next to your drawing. You should now have a two-dimensional perspective of the artificial interstitial habitat that you created on your slide.

5. Based upon the measurements you made of the meiofauna observed in this unit, insert *outlines* of the animals in the interstitial spaces that you have just drawn. Drawing the animals to the same scale as your sediment grains should reveal the amount of interstitial space available to the meiofauna.

Interpretation

1. Based upon your drawing of the interstitial habitat, could any of the animals ingest the sediment grains? If so, name them.

2. Based on what you know about the feeding habits of any animals named in question 1, would they ingest the grains?

3. If the motile meiofauna were chasing each other through the interstitial spaces, which animals would catch which other animals?

C. Meiofaunal Search

EXERCISE 1

Search-and-Sort Method for Gathering Meiofauna

A variety of methods are used to collect sedimentary material for meiofaunal analysis. These methods range from simple scoops to complex benthic grabs, depending on the depth of the water, the composition of the sediment, and the intent of the study. You may be exposed to other techniques used to extract meiofauna from sediments if field collections are returned to the laboratory for study.

This exercise emphasizes the laboratory search-and-sort component of the project. Sedimentary material from a well-conditioned marine aquarium may be more than adequate to test your skills.

PROCEDURE

1. Gather the items mentioned in General Laboratory Techniques step 1 (page 55).

2. With a small spatula, place a scoop of sedimentary material from a marine aquarium (or other source) in a Syracuse watch glass (or small petri plate).

3. Immediately scan the material with a dissecting microscope for any moving organisms. Can you provide some estimate of their abundance? For example, how many do you see in your microscopic field of view, relative to the sediment grains? You can quantitatively make such estimates by calculating the size of your microscopic field (see Appendix I).

NOTE: It is sometimes advantageous to stain the organisms. Water-soluble vital stains (crystal violet, neutral red, Janus green, safranin O, methylene blue, methylene green) may be administered in very dilute solution. They impart a distinctive color to living cytoplasm without interfering with metabolism. Alternatively, the sample or the extracted organisms (see below) may be immersed in 1% rose Bengal solution for 10 minutes. The solution is then carefully drained and replaced with water for 10 to 15 minutes to remove excess dye.

4. With a soft brush, splinter, or fine teasing needle, transfer any such material to a depression slide and observe the contents with a compound microscope.

5. Describe and sketch the different forms. How large are they? Are they sessile or motile forms? Can you identify the taxon to which the form belongs?

Interpretation

1. Prepare a report. Specify the source of your sample and the methods you used to gather your specimens. Provide drawings and descriptions requested, and answer the questions asked in the procedural steps.

REFERENCE

Feller, R. J., and R. M. Warwick. 1988. Energetics, pp. 181–196. In: R. P. Higgins and H. Thiel (eds.). *Introduction to the study of meiofauna*. Smithsonian Institute Press, Washington, D.C.

Aquatic Plants:
Macroalgae and Sea Grasses

OBJECTIVES

After completing this unit, you will be able to

- Understand the basic morphology of algae and sea grasses; and;

- Recognize the extent of morphological diversity of marine algae and sea grasses.

INTRODUCTION

Flowering plants **(division Anthophyta)** are dominant and successful on land, but barely represented in the sea. Instead, the oceans have seen a long and complex development of the morphologically simple, but physiologically and biochemically diverse algae (seaweeds). **Algae** are chlorophyll-bearing organisms that are **thalloid**—that is, they lack vascular tissues (true roots, stems, and leaves). So varied are the algae that a precise definition is difficult. Botanists recognize several distinct groups (divisions) based on differences in size, habitat, morphology, reproduction, pigments, physiology, and biochemistry.

Three divisions of algae make up most of the benthic flora: **Chlorophyta** (green algae), **Phaeophyta** (brown algae), and **Rhodophyta** (red algae). All three types of algae enjoy worldwide distribution, but their numbers and proportions vary according to habitat and climate. Discussions of algae often erroneously include a more primitive, chlorophyll-bearing group, the **Cyanobacteria** (blue-greens). Although blue-greens create a profound impact on the marine environment, they are not included in this unit.

This unit introduces the morphology of macroscopic marine algae and common flowering plants (anthophytes) that populate shallow subtidal regions.

A. Morphology and Diversity of Macroscopic Marine Algae

GENERAL PROCEDURE

1. Examine representative specimens of brown, green, and red algae. These may be live, or preserved, or herbarium specimens.

   a. If the specimen is **alive,** place it in a clean, opaque white container filled with seawater. If such containers are unavailable, place the specimen in a glass dish on a sheet of white paper. Be sure to keep the specimen moist.

   b. **Preserved** specimens must be kept moist. Place them in containers suitable for use with preservatives. You may wish to handle them with gloves, depending on the preservative used.

   c. If you are working with **herbarium specimens,** study them on a flat surface. Avoid bending the preparation. Keep them horizontal when moving them from one location to another.

2. Sketch the specimens and note their sizes, shapes, colors, and conspicuous structures, as directed in the exercises that follow. Use your notebook and drawing paper, as directed.

3. Examine prepared slides of thallus morphology.

4. Examine collections of algae that demonstrate diversity of form within the divisions.

EXERCISE 1

Division Phaeophyta

Brown algae are almost entirely marine and very diverse in form and structure. Large members of division Phaeophyta (dusky plants) exhibit remarkable adaptations to their environments. Most species live in the lowest intertidal or subtidal zones. Genera such as *Fucus* and *Laminaria* (Fig. 5.1a,b) form extensive intertidal and subtidal mats, respectively, on temperate to boreal waveswept rocky shores. Giant kelp (*Macrocystis pyrifera,* Fig. 5.1c) is massive (to 40 m long) and forms the subtidal kelp forests off the California coast. *Nereocystis* sp., or bull kelp (Fig. 5.1d), forms submarine kelp forests south of Alaska; it is commonly collected in shore drifts. Tropical regions have few species of brown algae, although some, such as *Sargassum* sp. (Fig. 5.1e), can occur in massive quantities in the open sea.

The characteristic brown coloration of this algae results from the presence of two accessory pigments, carotene and fucoxanthin, that mask chlorophyll. Food is stored as soluble carbohydrates.

ROCKWEED PROCEDURE

Examine a specimen of *Fucus* sp. (rockweed), a common temperate region marine alga of the rocky intertidal zone (Fig. 5.1a).

1. The **thallus** has a stabilizing **holdfast** that firmly attaches the thallus to rocks. Above the holdfast is a short stalklike part, the **stipe.**

2. The remainder of the plant is a flattened, dichotomously shaped **(Y-shape) blade.** The branches are thick and leathery, and they have a ridge running in their midline. The blade also contains **pneumatocysts** (air bladders) that keep the plant afloat at high tide.

3. At the tips of the branches are swollen, somewhat spherical and hollow **conceptacles.** The conceptacles are perforated with pores that lead into spherical cavities that contain **gametangia,** the structures that produce sperm (in males) or eggs (in females). If you are working with a fresh specimen, and if time permits, make an incision through a conceptacle and squeeze the contents onto a slide. Add a drop of seawater and examine the preparation with a compound microscope. Are any of the contents motile?

4. If you are observing a live specimen, feel the plants. Their slippery nature results from mucilaginous polysaccharides (complex sugars), common to this division. Expose a fresh specimen, or a portion of one, to air. Feel it every five minutes or so, and describe any changes that you observe.

5. Determine the ratio of surface area to volume for your specimen.

   a. Measure a 50-mm length along one of the Y-shaped branches, and mark the 0, 25-mm, and 50-mm points.

   b. At each of the three marks, measure the width and thickness of the blade. Calculate the average width and thickness of the 50-mm section.

   c. Calculate the surface area (SA), using the mean width and the 50-mm length. Remember to include both sides of the blade.

   d. Calculate the volume (*V*) of the blade using the mean width, thickness, and length.

   e. Calculate the SA:*V* ratio for the 50-mm portion.

Rockweed Interpretation

1. Did the blade of your specimen have a midrib? What is its function? Remember that algae lack vascular tissues.

2. Describe the problems algae might have because they live in the intertidal zone.

3. What adaptations does *Fucus* sp. have to aid survival in the intertidal zone? Consider the problems you described in question 2.

4. What advantage to rockweed does the flattened nature of the blade provide? What disadvantage does it create?

5. Brown algae are covered with a mucilaginous secretion. What happens to this coat when it dries? What is its function?

6. Refer to your calculations of surface area to volume ratio for your specimen of rockweed. Do you think your ratio is large? Is this advantageous to the alga? Explain.

KELP PROCEDURE

Examine specimens of bull kelp (*Nereocystis* sp.), oar weed (*Laminaria* sp.), or giant kelp (*Macrocystis* sp.) (Fig. 5.1b–d). Although you may observe a relatively small specimen, kelps normally grow to a much larger size than rockweeds and inhabit deeper regions of the coastal zone. Because kelps exhibit a high degree of anatomical specialization, the descriptions that follow may vary from species to species.

1. The **holdfast** consists of numerous branching structures called **haptera** that stabilize and attach the alga to the substrate. The haptera are not involved with nutrient absorption. The kelp may also have one or more flattened, elongated **blades** (or lamina) up to several meters in length, with narrow stalk-like **stipes.**

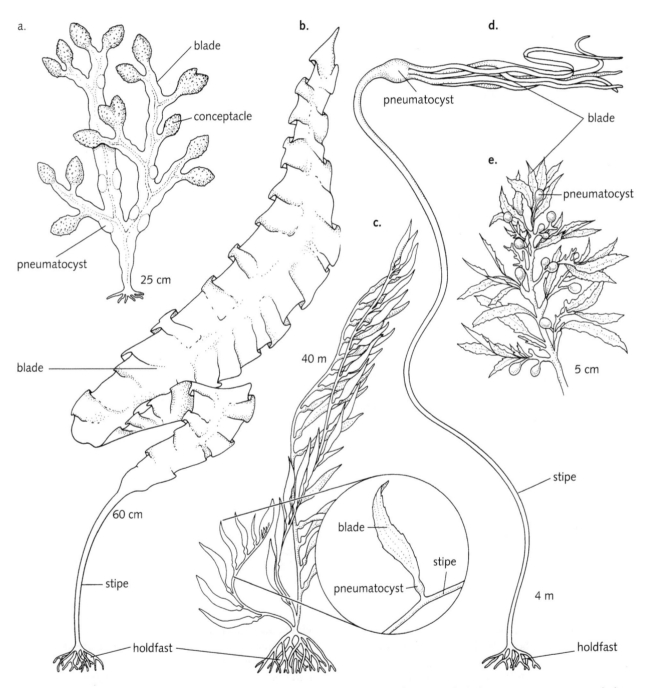

Figure 5.1 Division Phaeophyta, five forms of brown algae: (a) rockweed, *Fucus;* (b) kelp, *Laminaria;* (c) giant kelp, *Macrocystis;* (d) bull kelp, *Nereocystis;* (e) *Sargassum* weed. Not drawn to scale. Measurements indicate mature size of specimens.

2. Determine the surface area (SA) to volume *(V)* relationship of your specimen.

    a. Measure a 50-mm length near the widest part of a blade, and mark the 0, 25-mm, and 50-mm points.

    b. At each of the three marks, measure the width and thickness of the blade. Calculate the average width and thickness of the 50-mm section.

    c. Calculate the surface area (SA), using the mean width and the 50-mm length. Remember to include both sides of the blade.

    d. Calculate the volume *(V)* of the blade using the mean width, thickness, and length.

    e. Calculate the SA:*V* ratio for the 50-mm portion.

3. The body is completely covered by a mucilaginous layer. Expose a fresh specimen, or a portion of one, to air. Touch it every five minutes or so, and describe any changes that you observe.

4. Examine prepared slides of the blade of *Laminaria* sp. The several distinct regions of internal tissues have photosynthetic, food storage, supporting, and conducting functions. Can you find any external evidence of these regions?

    The blade is several cell layers thick (Fig. 5.2a). Note the symmetrical gradation of cell size from the outer **epidermis** through the **cortex** toward the interior core **(medulla)** of the blade. On its outer surface, the epidermis bears **sporangia**—organs that produce biflagellated zoospores.

    Just inside the epidermis is the cortex. The smaller (outer) cells of the cortex are photosynthetic (chloroplast-bearing) and form symmetrical layers on both sides of the thallus. The central medulla of the thallus consists of a mass of threadlike, or vacuolated cells, and mucilaginous secretions. How thick is the blade? What is the ratio of photosynthetic tissue to nonphotosynthetic tissue in the blade?

5. Although **pneumatocysts** are not present in *Laminaria* sp., they are conspicuous in the kelps. In *Macrocystis* sp., one air bladder appears at the base of each blade. Bull kelp has a single massive pneumatocyst at the upper end of a long, slender stipe. Multiple ribbonlike blades extend from the bulblike float.

Kelp Interpretation

    1. Compare the surface area to volume ratio of kelps with that of rockweed.

SARGASSUM PROCEDURE

Examine a piece of Sargassum weed, *Sargassum* sp. (Fig. 5.1e), a floating brown alga of tropical waters. It begins its life cycle as an attached thallus but soon

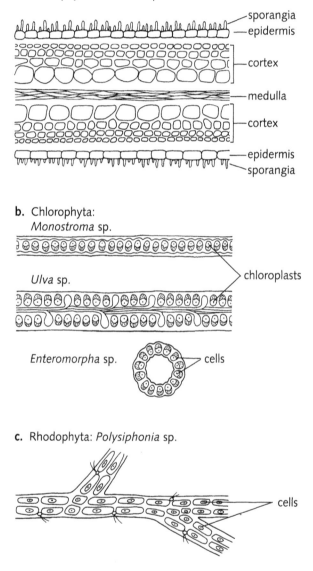

a. Phaeophyta: *Laminaria* sp.

sporangia
epidermis
cortex
medulla
cortex
epidermis
sporangia

b. Chlorophyta: *Monostroma* sp.

Ulva sp.

chloroplasts

Enteromorpha sp.

cells

c. Rhodophyta: *Polysiphonia* sp.

cells

Figure 5.2 Microscopic anatomy of thalli (blades) of algae. Transverse sections of (a) brown (*Laminaria* sp.), and (b) green algae (*Monostroma* sp., *Ulva* sp., *Enteromorpha* sp.). (c) Single, branched filament of red alga (*Polysiphonia* sp.) shows tier arrangement of cells.

floats and continues to multiply by fragmentation. Among the most highly differentiated of the algae, these plants extend for hundreds and thousands of hectares over the sea's surface, especially in the Sargasso Sea. Much of the weed in the Sargasso Sea originates in the Gulf of Mexico and is carried into the area around Bermuda by the Gulf Stream and its eddies.

Observe the stemlike **stipe,** the serrated leaflike **blades** (with veins), and solitary, but numerous **pneumatocysts.** The genus name *Sargassum* is derived from the Portuguese word for grape, and refers to the gas-filled bladders.

Sargassum Interpretation

1. Would you say that the thallus of *Sargassum* is more or less complex than that of *Fucus* or *Laminaria?* Defend your argument.

EXERCISE 2
Division Chlorophyta

The green algae (Fig. 5.3) contain the same photosynthetic pigments (chlorophylls *a* and *b*) as terrestrial plants; the food reserve is starch. The division Chlorophyta (green plants) plays a relatively minor role in the sea, compared with their overwhelming dominance in freshwater. On the other hand, certain coastal species (*Enteromorpha* sp., *Ulva* sp.; Fig. 5.3a,c) can occur in great abundance, but in stands of low diversity. Chlorophytes occur primarily in the upper 10 m of the water column, reaching maximum population densities in the upper 3 m. Some membranous forms (*Monostroma* sp. and *Ulva* sp.) may be only one or two cell layers thick (Fig. 5.2b,c). Others are tubular (*Enteromorpha* sp., Figs. 5.2b and 5.3a), or filamentous and branched (*Cladophora* sp., Fig. 5.3d). The branched thallus of *Codium* sp. (Fig. 5.3e) has the consistency of firm gelatin. The greatest evolutionary diversification of the group occurs in the tropics, where prominent and variable assemblages of forms occur, many of them with calcareous coverings, such as *Udotea* sp., *Penicillus* sp., and *Halimeda* sp. (Fig. 5.3f–h). Some species, such as *Acetabularia* sp., and *Valonia* sp. exhibit extraordinary morphology (Fig. 5.3i,j). *Acetabularia* sp. is composed of a single cell, and has a gracefully stalked, saucer-like disc; *Valonia* sp. is balloonlike, with globular, sausage-shaped, or club-shaped thalli.

PROCEDURE

Examine a specimen of *Ulva lactuca* (sea lettuce) (Fig. 5.3), a cosmopolitan (worldwide) intertidal form.

1. The **blades** of the **thallus** are membranous and relatively broad and may be 40 cm long. The blades are more substantial than those of the closely related genus *Monostroma*—they are two cells thick rather than one (Fig. 5.2b).

   Sea lettuce is relatively tough and reasonably well-adapted for survival in the intertidal zone. *Ulva lactuca* typically grows attached to a substrate by a **holdfast** with rhizoid-or finger-like extensions; it is most often seen in free-floating masses, however.

2. Make a wet mount (see Unit 4 for instructions) of a small piece of the blade of *Ulva* sp., and examine it under low power on the compound microscope. Can you determine how many cell layers are present? If not, try switching to high-dry magnification. Focus on the uppermost surface. Do you see any hint of cells beneath the surface layer? Adjust the microscope so that these layers come into focus. Notice that your specimen is transparent. Could you make the same observations on the *Laminaria* sp. blade?

3. Measure the width of a 50-mm length of blade at its widest point. Calculate the surface area and volume for that segment.

4. Examine other specimens of green algae that may be available for observation. Compare their sizes, shapes, and textures.

Interpretation

1. What is the size of your specimen? How does the ratio of its surface area to its volume compare with that of brown algae you may have observed?

2. Does your specimen have a holdfast? Should it have one?

3. Would you expect *Ulva* sp. to be found in subtidal zones? Explain. You may want to refer to Unit 7 for information on plant pigments.

4. Would you expect *Ulva* sp. to grow in areas of rapid current flow? Explain.

Figure 5.3 Division Chlorophyta, examples of temperate and tropical green algae. Sizes of mature individuals indicated.

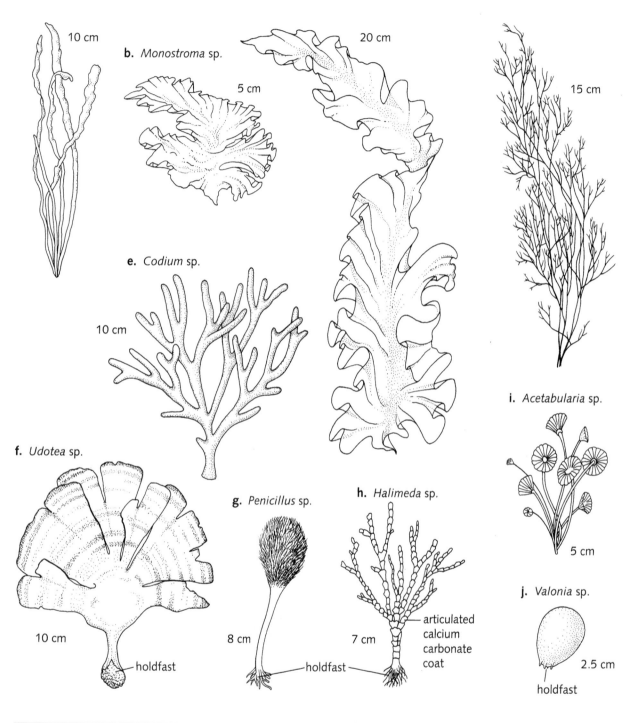

a. *Enteromorpha* sp.

10 cm

b. *Monostroma* sp.

5 cm

c. *Ulva* sp.

20 cm

d. *Cladophora* sp.

15 cm

e. *Codium* sp.

10 cm

f. *Udotea* sp.

10 cm

holdfast

g. *Penicillus* sp.

8 cm

holdfast

h. *Halimeda* sp.

7 cm

articulated
calcium
carbonate
coat

holdfast

i. *Acetabularia* sp.

5 cm

j. *Valonia* sp.

2.5 cm

holdfast

E X E R C I S E   3
Division Rhodophyta

Red algae comprise the largest and most diversified assemblage of marine algae. The members of division Rhodophyta (red plants) are the most numerically dominant, macroscopic rock-covering algae occurring in virtually every part of the world. Red algae vary widely in size and complexity, from thin films to complex filamentous and membranous forms growing to heights approaching one meter (Fig. 5.4). Most are exclusively benthic, attached organisms and occupy all latitudes. Some live in depths exceeding 35 m.

Figure 5.4 Division Rhodophyta, diverse forms of red algae. Sizes of mature individuals indicated.

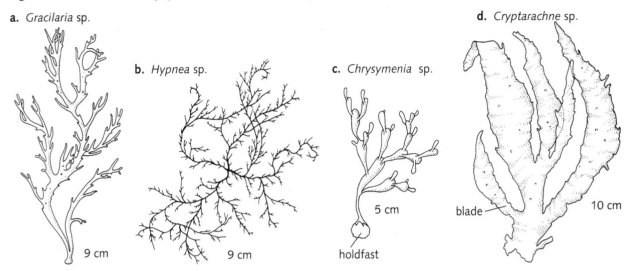

a. *Gracilaria* sp.

b. *Hypnea* sp.

c. *Chrysymenia* sp.

d. *Cryptarachne* sp.

9 cm

9 cm

5 cm

holdfast

blade

10 cm

Figure 5.5 Division Rhodophyta, calcium carbonate secreting red algae. Sizes indicate mature specimens.

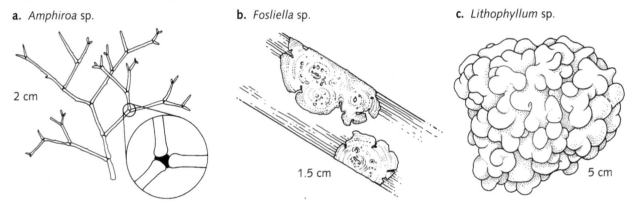

a. *Amphiroa* sp.

b. *Fosliella* sp.

c. *Lithophyllum* sp.

2 cm

1.5 cm

5 cm

Specialized pigments (chlorophylls *a* and *d,* phycoerythrin and phycocyanin), support photosynthesis in deep water. Many red algae are epiphytic; some are parasitic. Some forms—the intertidal coralline (coral-like) algae (Fig. 5.5)—secrete calcium carbonate ($CaCO_3$). **Articulated** species, such as *Amphiroa* sp. (Fig. 5.5a), form a skeleton on the blades; **crustose** species (*Fosliella* sp., Fig. 5.5b) form encrusting plate-like layers on rocks and shells. Crustose algae blanket more than one-third of fringing reef surfaces and act as the major builders and consolidators of reef material, surpassing the reef contributions of all other organisms.

PROCEDURE

1. Examine any available specimens. Pay particular attention to their varied forms, and compare them with specimens from the other divisions. Note whether the blades are flattened or cylindrical, simple or branched. Do they have **pneumatocysts?**

2. Make a wet mount (see Unit 4) of a small portion of a filamentous red alga, or obtain a prepared slide of *Polysiphonia* sp. Examine your preparation under low power with a compound microscope. Select a small branch of the alga, and switch to high-dry magnification. Can you determine the shape of the branches (flat, irregular, cylindrical)? How many cell layers does the branch contain? Make a transect of one of the filaments and prepare another wet mount. Compare your observations with the illustration of *Polysiphonia* sp. in Figure 5.2c.

Interpretation

1. What advantage does a calcareous skeleton provide to algae?

2. What pigments give this group its color?

3. Is it logical to expect this group to use red light for photosynthesis? (Refer to Unit 7 for information on pigments.)

4. It was easy to determine proportions of surface area to volume in the blades of brown and green algae. How does the surface area of the red algae differ?

5. Can you explain why red algae lack pneumatocysts?

Summary Interpretation of Algal Morphology and Diversity

1. Construct a table that summarizes what you observed and what you learned from reading and lecture. List the divisions observed in the first column. Label successive columns as species, common name, pigments, thallus (shape, texture), pneumatocyst (+/−), conceptacle (+/−), surface area:volume, and any other characteristics.

2. The effects of water motion are important in the design of seaweed thalli. Plants regularly exposed to wave action have more structural tissue than species living in calmer waters. You can observe this characteristic by examining different species that exist in different wave energy habitats. Variations can also be seen among individuals within the same species if they are collected from different habitats. The increase in structural tissue reduces the ratio of photosynthetic tissue to nonphotosynthetic tissue.

   a. Did you observe major differences in this ratio in this exercise? Try to relate your observations with the known habitat of the observed species.

3. Thalli possess various features that reduce resistance to water flow: (a) a shape that ensures horizontal posture with the current, (b) a smooth surface that reduces skin friction, and (c) production of mucilage that reduces drag.

   b. Did you observe any of these traits? Relate your observations with the known habitat.

4. Describe how the morphology and other characters listed in your table are correlated with known habitat of the species. Is there an adaptive significance to the differences? Consider, for example, the following questions:

   c. Which species are more robust?

   d. Do they all have mucilaginous coats?

   e. How does shape relate to photosynthetic efficiency?

   f. How is shape hazardous to survival?

5. How do algae acquire mineral nutrients?

6. Where is the principal growing region of algae?

B. Morphology and Diversity of Sea Grasses

Although flowering plants (division Anthophyta) have invaded the sea only in relatively recent geologic time, certain species have been highly effective colonizers of the coastal habitat. The extensive beds of sea grasses in the sublittoral (subtidal) zone offer a clear example of this success in colonization. The majority of these are tropical in distribution, but some—for example, *Zostera* sp. (Fig. 5.6)—are typically temperate.

Marine vascular plants must: (a) tolerate immersion in a saline medium; (b) be able to grow when totally submerged; (c) have an anchoring system able to withstand tidal currents and wave action; (d) pollinate underwater; and (e) compete successfully in the marine environment.

EXERCISE 1

Examining Representative Sea Grasses

Examine available specimens of sea grasses as you read through the following general account (modified from Phillips and Menez, 1988) of the structural features of leaves and rhizome-root systems. This account is followed by morphological descriptions and life histories of individual representative species.

Make appropriate notes and sketches in your notebook as you work.

1. Generalized Structure of Leaf and Rhizome-root

   Except for *Syringodium* spp. (Fig. 5.9c,d), whose leaves are cylindrical, all genera of sea grasses have relatively thin, bladelike, and flattened leaves with a high surface-to-volume ratio. This characteristic provides for maximum diffusion of gases and nutrients between the blades and seawater, and maximum photosynthetic surface.

   The **leaves** lack stomata, but diffusion of gas and nutrients occurs through the thin **cuticle.** The flexible blades lack supporting tissue, which allows them to be swept by currents. The blades exert a frictional drag on the water, thereby reducing current velocity and sediment erosion within the meadow, and increasing particulate organic sedimentation.

   The morphology of the **rhizomes** and roots complements anchorage and nutrient absorption. Rhizomes have supporting tissue fibers that provide structural rigidity to the below-ground system. An extensive system of air chambers **(lacunae)** in the roots connects with that in the leaves. This system facilitates transport of oxygen to the roots, and allows the roots to grow in anoxic environments. All sea grasses produce **root hairs,** whose abundance varies with species.

Sea grass plant morphology varies seasonally, with depth, substrate texture, and substrate nutrient levels. With eel grass, for example, leaf-to-rhizome/root ratios vary along a sediment texture gradient where more roots are formed in mud than in sand. Winter ratios favor a dominance of roots because of a decline in leaf biomass. Roots also have greater biomass in the upper intertidal zone than in the lower intertidal zone. Shoot density may be 3–4 times greater in summer than in winter. Leaf dimensions are shorter and narrower in winter (than in summer), and in the intertidal zone (than in the subtidal zone).

2. Representative Species

a. Eel grass, *Zostera marina* The most widely distributed sea grass in North America (Atlantic and Pacific coasts) is eel grass (Fig. 5.6). It generally occurs in sheltered, shallow lagoons and bays, on mixed mud and sand substrate. Eel grass supports a great variety of marine animal life and serves as the staple winter food for waterfowl.

Eel grass **leaves** are linear, up to 2 m long, and 1.5–12 mm wide. Five to eleven main veins run the length of the leaf, growing in clumps from a stoloniferous rhizome.

Upon germination of the **seed** (Fig. 5.6b), the **radicle** (new root) produces abundant root hairs, while the **shoot** elongates through several centimeters of muddy substrate. **Adventitious roots** (accessory roots)

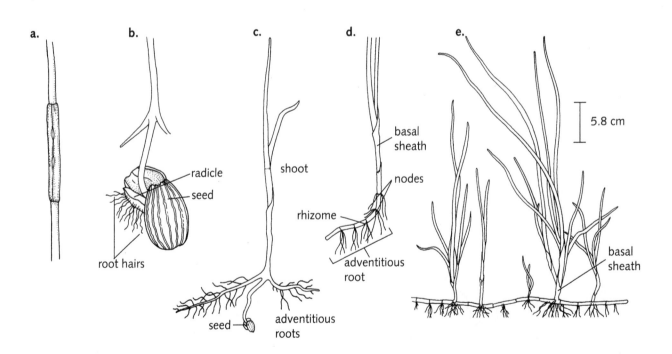

Figure 5.6 Division Anthophyta, *Zostera marina,* eel grass: (a) fertile shoot with ripening fruits partially evident; (b) seed and lower part of young seedling showing protruding radicle with root hairs, and first pair of secondary (adventitious) roots at first node; (c) stage in development of seedlings; (d) lower part of older plant showing rhizome and roots growing from nodes; (e) portion of a natural stand.

develop early from opposite sides of the first **node** (the point on the shoot from which leaves or branching occurs) (Fig. 5.6c). Further growth involves development of a **rhizome** (horizontal, underground stem) by elongation of the **internode** (space between two successive nodes) and loss of the first leaves. Adventitious roots develop from each node, and from time to time lateral buds sprout at the nodes. One to two years may be required for the plant to reach stage E (Fig. 5.6d). The **fertile shoots** (Fig. 5.6a) perish at the end of the season, while the roots persist but fragment.

Flowers consist of either a pistil or a stamen. They occur alternately in two rows. From 500 to 1000 seeds may be produced on a luxuriant plant during one season; fish and waterfowl eat most of them, however. A few become favorably buried in the substrate by being carried into cavities and tunnels of mud-burrowing benthic animals.

b. Turtle grass, *Thalassia testudinum* Caribbean turtle grass (Fig. 5.7) is the most abundant sea grass in the tropical western Atlantic region, appearing from Florida into the Caribbean and along the coast of the Gulf of Mexico. It occurs in a variety of loose substrates—from mud to coral sand to broken shell, wherever relatively calm water holds these materials in position. A dense intertidal growth can occur in quiet lagoons, while in open water beds may be found at depths of up to 30 m. The plants tolerate salinity variations from 10 o/oo to 48 o/oo, with optimum range of 25–38 o/oo.

The **leaves** are dark green and blunt at the tip (Fig. 5.7b); they are usually covered by epiphytes and calcareous deposits. One main **vein** and several smaller veins run the length of the leaf. The leaf clusters are **sheathed** at the base. The **rhizome** is hairy, with coarse, fibrous **adventitious roots.**

Thalassia develops its erect, leafy shoots from a creeping rhizome buried 5–10 cm in the substrate. New erect shoots consist of a short stem with a small group of four to five exposed leaves (Fig. 5.7b). The stem is dioecious, bearing staminate and pistillate flowers (Fig. 5.7a). Flowering occurs from May to July in Florida.

c. *Halophila decipiens* This species (Fig. 5.8) occurs in sheltered sites from the lower intertidal zone to 85 m deep. It is widely distributed in tropical parts of the Pacific and western Atlantic Oceans.

This genus has flattened, but ovate **leaves.** In *H. decipiens,* the serrated leaves are 10–25 mm long and 3–6 mm wide.

d. Surf grass, *Phyllospadix torreyi* This species (Fig. 5.9a,b) occurs from British Columbia to Baja California on surf-beaten rocky coasts, but is not found on the most exposed sites. It colonizes via rhizomes growing into open spaces or by seed dispersal. The latter method can occur only if macroalgae, such as *Ulva* sp., are present to serve as a substrate for seed attachment.

Leaf blades are 0.5–2 m long and 0.5–1.5 mm wide, and enclosed in a **sheath.** The older parts of the rhizome are covered with pale yellow to gray fibers.

e. *Syringodium filiforme* (= *Cymodocea filiforme)* This species (Fig. 5.9c) is confined to the subtidal zone, and often occurs mixed with *Thalassia testudinum* from low tide to 10 m. The species occurs throughout the western tropical Atlantic, in the Gulf of Mexico, along the east coast of Florida, and in Bermuda.

The **leaf** blades are cylindrical, very long (1–30 cm), and narrow (0.8–2 mm).

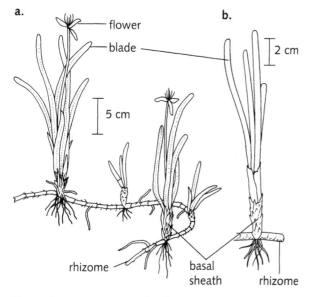

Figure 5.7 Division Anthophyta, turtle grass, *Thalassia testudinum:* (a) portion of natural stand; (b) enlarged portion of one set of shoots.

Figure 5.8 Division Anthophyta: (a) mature stand of *Halophila decipiens* showing male and female flowers; (b) enlargement of leaf and serrated edge of leaf.

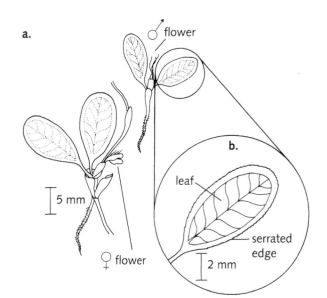

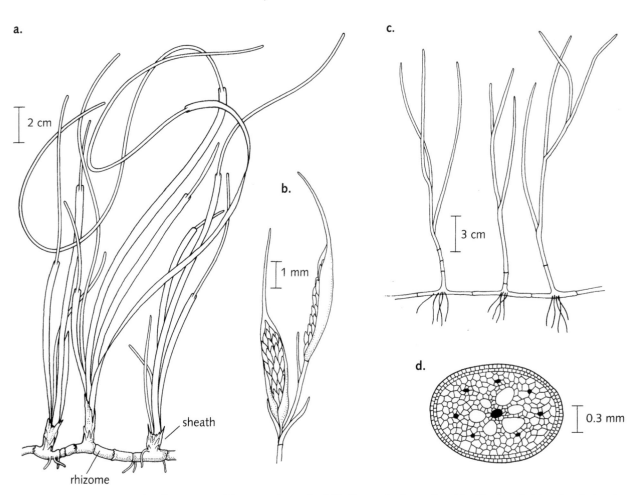

Figure 5.9 Division Anthophyta: (a and b) natural stand of *Phyllospadix torreyi* and fertile shoot; (c) natural stand of *Syringodium filiforme;* (d) section through blade.

Epibiota of Algae and Sea Grasses

OBJECTIVES

After completing this unit, you will be able to

- Describe species composition of epibionts on the blades and/or holdfasts of macroalgae and sea grasses;

- Determine the extent to which epibionts cover the blades of macroalgae or sea grasses;

- Compare coverage and species composition of epibionts on different-age algal fronds, sea grass blades, or holdfasts.

INTRODUCTION

Rapid and dense growth of macroalgae (divisions Chlorophyta, Rhodophyta, and Phaeophyta) and sea grasses (division Anthophyta) provides food, substrate, and shelter for a variety of marine organisms. Macroalgal and sea grass systems, with their associated flora and fauna, comprise complex epibenthic communities distinct from others in the immediate area.

The surface of macroalgae and sea grasses provides a substrate for the settlement of microorganisms, such as bacteria, cyanobacteria, protistans, and diatoms. These microorganisms are grazed by progressively larger organisms, including meiofauna and macrofauna. The microbiota also establish a base for the growth and development of an assortment of other species (algae included) that settle and colonize any available exposed substrate. This **epibiotic community** is often so abundant that its biomass exceeds that of the sea grass blade. Members of this community are called **epibionts;** animal members are known as **epifauna,** while plant members are called **epiphyta.**

The complex holdfasts of brown algae support their own unique communities. They provide temporary refuge for motile benthos, such as brittle stars and polychaete worms, and relatively secure anchorage sites for mussels and sea squirts. A large contingent of encrusting and filamentous forms, such as sponges and bryozoans, exists on the surface area of the brown algae.

PROCEDURE

EXERCISE 1
Diversity of Motile Epifauna

1. Select an algal or sea grass plant, and identify it. Work in teams as directed. If several specimens of the same species are studied by your class, they will be designated by letter (for example, A–F). Record this information in the spaces provided on the Report Forms A–E.

2. Sketch the plant and conspicuous epibiota on Report Form A. Label the sketch as completely as you can. Provide a scale bar to indicate size. Add to the sketch as you progress through the exercise.

3. Hold the specimen over a sieve and wash it with seawater to remove *motile epifauna*.
   NOTE: To trap meiofauna, use a sieve with 63-μm mesh size; to capture macrofauna only, use a 0.5-mm mesh screen.

4. Transfer the specimens to seawater in a petri plate, Syracuse dish, or depression slide.

5. Blot the plant specimen dry and determine the area of the blade.

   a. Trace outlines of the blades on a piece of metric graph paper (with a grid showing 10 mm to the cm).

b. Count the number of mm squares within the outlines. Remember to include both surfaces of the specimen.

c. Record the area values (as cm²) on Report Form B.

6. Identify the *motile epifauna* to the lowest possible taxon.

a. Use the Key to Macroscopic Coastal Marine Invertebrates in Unit 2.

b. Refer to Table 11.2 (checklist of commonly occurring organisms in coastal fouling communities) in Unit 11.

c. Have your instructor confirm your identifications.

d. Count the specimens of each species. Record this information on Report Form C, and share your data with the class (see Report Form D).

7. Calculate diversity indices (H and D) for your plant. (Equations and a sample set of calculations may be found in Appendix II.) Make your calculations on Report Form E, and record your indices at the bottom of Form C.

8. Express the abundance of motile epifauna as number per 100 cm² of blade area (both surfaces). Record the information in the third column of Report Form C.

9. Share your identifications and counts of individuals with the class. Record the class data on Report Form D. Use additional sheets if there are more than six plants in your class.

10. Calculate the Index of Similarity for all pair combinations of "plants" examined by the class, and record your indices on Report Form E. Treat each sample as a separate plant. (Sample calculations are provided in Appendix II.)

E X E R C I S E 2

Coverage of Sessile Epibiota

1. Scan both surfaces of the plant specimen with a dissecting microscope. Locate the sessile epifauna and epiphyta.
NOTE: If the plant has an easily recognized upper and lower side (as in a sea grass blade),

distinguish between the sides, and record your observations accordingly.

2. Identify the fauna to the lowest possible taxon.

a. Use the Key to Macroscopic Coastal Marine Invertebrates in Unit 2. Refer to Table 11.1 in Unit 11.

b. Confirm your identifications with the instructor, and record them on Report Forms F and G.

c. Classify vegetative growth as encrusting or filamentous "epiphyta."

d. Classify animal growths as eggs, juvenile stages, or adult forms.

3. Determine the surface area of the epibionts.

a. Sketch their coverage on your outline drawings (Exercise 1, step 5).

b. Count the number of squares occupied by each species.

c. Record the data on Report Forms F and G.

4. Use Report Form H and combine the data from the upper (Form F) and lower (Form G) surfaces of your plant.
NOTE: This step involves combining data from species common to both surfaces. Certain species may be present on either the upper or lower surface, hence Form H may have more species listed than either Form F or G.

5. Calculate percentage coverage of the sessile forms, based on the combined values on Form H. Rank the species according to coverage, with 1 = greatest percentage.
NOTE: Coverage can exceed 100% if organisms occur in layers.

6. Use Report Form I to compile class data of species observed. The count for each species (n_i) will consist of the number of individuals or the number of colonies (for encrusting and filamentous forms).

7. Calculate diversity indices for all specimens (for example, A–F) and similarity indices for the pairs of sessile epibiotic communities. Refer to procedural Exercise 1, steps 7 and 10, and to Appendix II for examples. Record the information on Forms I and J.

EXERCISE 3

Density and Coverage of Epibiota in Relation to Age of Plant

1. Select specimens (blades or entire plants) of different age (by size).

2. Perform the analyses described in Exercises 1 and 2 for each plant specimen. Use separate sets of data forms for each specimen.

EXERCISE 4

Holdfast Community Analysis

This community may be treated in depth as described in Unit 11 (Fouling Communities), or similar to Exercise 1 (Motile Epifauna) in this unit. Report Forms A, C, D, and E are suitable for holdfast analysis. You may prefer to determine holdfast volume (by volume displacement method), and record epibiont abundance in terms of volume rather than area.

Epibiota Interpretation

1. Rank the species according to: (a) density (number/ 100 cm²) of motile epifauna (Form C); (b) numerical abundance of holdfast inhabitants; and (c) percentage coverage of sessile forms (Form H). Compare your observations with published data of other studies. See, for example, papers on eel grass (Marsh, 1973), turtle grass (Lewis and Hollingworth, 1982; Hall and Bell, 1993), and *Sargassum* weed (Butler et al., 1983).

2. Compare your numerical abundance and species diversity values with those obtained by other members of your class. How can you account for similarities and differences among the samples?

3. Are there any differences in coverage by the various epibionts? Is coverage confined to only one surface of the blade? Is coverage heavier on one side or the other? Do the various species show any preference in location of settlement? Attempt to explain any such observed differences in coverage.

4. What is the relative proportion of eggs and egg cases to juvenile stages to adults in your epibiotic community? Consider the season during which your sample was collected. Would collections from other times of the year exhibit any difference in the population composition?

5. What can you conclude from your examination of different age/size specimens?

REFERENCES

Butler, J. N., B. F. Morris, J. Cadwallader, and A. W. Storer. 1983. *Studies of* Sargassum *and the* Sargassum *community*. Bermuda Biol. Station Spec. Publ. No. 22, 307 pp.

Hall, M. O., and S. S. Bell. 1993. Meiofauna on the sea grass *Thalassia testudinum:* population characteristics of harpacticoid copepods and associations with algal epiphytes. *Mar. Biol.*, 116:137–146.

Lewis, J. B., and C. E. Hollingworth. 1982. Leaf epifauna of the sea grass *Thalassia testudinum. Mar. Biol.*, 71:41–49.

Marsh, G. A. 1973. The *Zostera* epifaunal community in the York River, Virginia. *Chesapeake Sci.*, 14:87–97.

EPIBIOTA REPORT FORM A

COLLECTION DATA

Date _____ Location _____ Your Station No. _____

Specimen (grass or alga):   Common Name _____

Specimen (A B C D E F) Species _____

SKETCH OF SPECIMEN:

Note size, position of organisms, location and direction of attachment, encrusting species, extent of coverage. Include a scale bar to indicate size.

NAME _____ SECTION _____ DATE _____

EPIBIOTA REPORT FORM B

Plant Species _____ Specimen (A B C D E F)

Blade Area   of specimen, as determined from outline sketches on graph paper.

Leaf or Blade	Area (cm^2)
1	
2	
3	
4	
5	
6	
7	
8	
9	
10	
Total	

Total Surface Area (SA), both surfaces included = _____ cm^2

EPIBIOTA REPORT FORM C

Plant Species _____ Specimen (A B C D E F)

MOTILE EPIFAUNA COMMUNITY OR HOLDFAST COMMUNITY

Both surfaces of your plant; n_i = count of individual animal species; N_p = sum of n_i for your plant.

Epifaunal Species	n_i	n_i/100 cm²	Rank*	Feeding Type**
1				
2				
3				
4				
5				
6				
7				
8				
9				
10				
11				
12				
13				
14				
15				

Totals:

Species = $\qquad$ $N_p =$ $\qquad$ N_p/100 cm² =

$\quad$ H = $\qquad\qquad$ D =

*Rank the species according to density, highest to lowest.
**Based on your reading, indicate if the species are predators (P), suspension feeders (S), or grazers (G).

NAME _____ SECTION _____ DATE _____

EPIBIOTA REPORT FORM D (CLASS DATA)

Plant Species _____ No. Plants = _____

DIVERSITY

MOTILE EPIFAUNA OR HOLDFAST; n_i = count of individuals in each species; N_i = sum of n_i on all plants; N_p = sum of n_i for each plant; N_t = grand total of all species on all plants.

| Species | Plants | | | | | | Species Total |
	A	B	C	D	E	F	
	n_i	n_i	n_i	n_i	n_i	n_i	N_i
1							
2							
3							
4							
5							
6							
7							
8							
9							
10							
11							
12							
13							
14							
15							

Totals: N_p = _____ + _____ + _____ + _____ + _____ + _____ = N_t _____

Species =
H =
D =

EPIBIOTA REPORT FORM E

Plant Species _____

MOTILE EPIFAUNA COMMUNITY OR HOLDFAST COMMUNITY

Diversity Indices: Calculations based upon your plant specimen; data from Report Form C. See Appendix II for explanation and sample calculations.

Shannon–Wiener $(H) = - \Sigma (n_i/N_p) \ln (n_i/N_p)$
$$H =$$

Simpson's $(D) = 1 - \Sigma (n_i/N_p)^2$
$$D =$$

n_i = count of individuals in each species; N_p = sum of n_i for each plant (sample); ln = natural logarithm.

Similarity Index: $SI = 2C/(A + B)$

C = number species in common on both plants A and B
A, B = number species on plants A and B, respectively

Table of SI for six plants. Expand table as needed. See Appendix II for further explanation.

Plant	A	B	C	D	E	F
A	———					
B	———	———				
C	———	———	———			
D	———	———	———	———		
E	———	———	———	———	———	
F	———	———	———	———	———	———

EPIBIOTA REPORT FORM F

Plant Species _____ Specimen (A B C D E F)

SESSILE EPIBIOTA   (*upper surface*—your plant).   Indicate form as filamentous (fil), encrusting (enc), adult (ad), juvenile (juv), or egg/egg case.

Species	Form	Area (cm^2)
1		
2		
3		
4		
5		
6		
7		
8		
9		
10		
11		
12		
13		
14		
15		
Total		

No. Species (Spp) =

EPIBIOTA REPORT FORM G

Plant Species _____ Specimen (A B C D E F)

SESSILE EPIBIOTA   *(lower surface—your plant).*   Indicate form as filamentous (fil), encrusting (enc), adult (ad), juvenile (juv), or egg/egg case.

Species	Form	Area (cm²)
1		
2		
3		
4		
5		
6		
7		
8		
9		
10		
11		
12		
13		
14		
15		
Total		
No. Species (Spp) =		

EPIBIOTA REPORT FORM H

Plant Species _____ Specimen (A B C D E F)

SESSILE EPIBIOTA   *(percentage coverage of total blade area—your plant).* Data from Report Forms F and G.

Species	Area (cm²)				Rank*
	Top	Bottom	Total	%Coverage	
1					
2					
3					
4					
5					
6					
7					
8					
9					
10					
11					
12					
13					
14					
15					
16					
17					
18					
Total					

*Rank species according to blade coverage, highest (= 1) to lowest

EPIBIOTA REPORT FORM I (CLASS DATA)

Plant Species _____ No. Plants = _____

SESSILE EPIBIOTA DIVERSITY n_i = count of individuals or colonies for each species; N_i = sum of n_i on all plants; N_p = sum of n_i for each plant; N_t = grand total of all species on all plants.

	Plants						Species Total
	A	B	C	D	E	F	
Species	n_i	n_i	n_i	n_i	n_i	n_i	N_i
1							
2							
3							
4							
5							
6							
7							
8							
9							
10							
11							
12							
13							
14							
15							

Totals: N_p = _____ +_____ +_____ +_____ +_____ +_____ = N_t _____

Species =
H =
D =

EPIBIOTA REPORT FORM J

Plant Species _____

SESSILE EPIBIOTA COMMUNITY

Diversity Indices: Calculations based upon your plant specimen; data from Report Form C. See Appendix II for explanation and sample calculations.

Shannon–Wiener $(H) = - \Sigma (n_i/N_p) \ln (n_i/N_p)$

$$H =$$

Simpson's $(D) = 1 - \Sigma (n_i/N_p)^2$

$$D =$$

n_i = count of individuals in each species; N_p = sum of n_i for each plant (sample); ln = natural logarithm.

Similarity Index: $SI = 2C/(A + B)$

C = number species in common on both plants A and B
A, B = number species on plants A and B, respectively

Table of SI for six plants. Expand table as needed. See Appendix II for further explanation.

Plant	A	B	C	D	E	F
1	_____					
2	_____	_____				
3	_____	_____	_____			
4	_____	_____	_____	_____		
5	_____	_____	_____	_____	_____	
6	_____	_____	_____	_____	_____	_____

Photosynthetic Pigments of Marine Macroalgae

OBJECTIVES

After completing this unit, you will be able to

- Understand the process of light energy absorption by chlorophyll in macroalgae;

- Extract chlorophyll from algal tissue;

- Determine absorption spectra of the extracted chlorophyll; and

- Separate and identify extracted pigments.

INTRODUCTION

With few exceptions, the capture of solar energy in photosynthetic reactions of chlorophyll-bearing plants **(autotrophs)** establishes the basis **(primary productivity)** for the flow of energy **(trophic spectrum)** of food webs. Autotrophs build complex organic molecules that contribute to the functions of all cells. Thus, the ultimate source of energy expended by most organisms is the converted energy of sunlight that is incorporated in the newly synthesized molecules during photosynthesis.

The process, which involves a series of sequential reactions, can be summarized by the chemical equation:

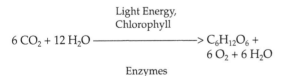

$$6\,CO_2 + 12\,H_2O \xrightarrow[\text{Enzymes}]{\text{Light Energy, Chlorophyll}} C_6H_{12}O_6 + 6\,O_2 + 6\,H_2O$$

Light energy is absorbed by the chlorophyll molecule and used to split water into hydrogen atoms and oxygen. Energy released during that process is incorporated into energy-rich ATP (adenosine triphosphate). The energy in ATP is used in a series of reactions that convert CO_2 to sugar ($C_6H_{12}O_6$). The sugar may be used in the plant's metabolism, or it may be stored as a more complex carbohydrate.

Light

Photosynthesis is possible only when light reaching the algal cells exceeds a certain intensity. This restriction limits the depth range of phytoplankton to the uppermost layers of the ocean, and the depth range of benthic macroalgae to shallow coastal regions. The depth to which light penetrates seawater—and thus the depth at which primary productivity can occur—depends upon several factors: latitude; season of the year; light reflection from the surface of the water; water transparency; absorption of light by water; and the wavelength of light.

The portion of light that enters the water is altered with depth because water does not absorb all wavelengths of light equally. Solar radiation includes all visible colors from violet to red (wavelengths from 400 nm to 700 nm; one nanometer = 10^{-9} m)(Fig. 7.1). Violet and red are absorbed within the first few meters; green and blue penetrate deeper.

Differential absorption of light intensity and wavelength by seawater contribute to vertical zonation of benthic macroalgae. In response to these environmental factors, benthic algae have evolved various morphological and biochemical adaptations. Examples of morphological adaptations include species with elongated thalli (*Laminaria* sp.) or flotation organs (*Macrocystis* sp.) (see Unit 5). Such species are anchored subtidally, but their thalli and blades reach to the sea's surface. Biochemical adaptations involve the presence of accessory pigments that are known to absorb light at intensities and wavelengths (green, violet, blue) present beneath the surface. Thus, certain species may exist in deeper or more turbid waters than others.

Pigments

There are three broad chemical categories of algal pigments: **chlorophylls, carotenoids** (both soluble in organic solvents), and water-soluble **phyco-bilins.** Each algal division contains special pigments or mixtures not found in others (Table 7.1).

Photosynthesis in algae is always associated with the presence of *chlorophyll a*, although *chlorophyll b* or *c* may also be present. *Chlorophyll b* is a light-harvesting pigment, transferring absorbed energy to *chlorophyll a*. *Chlorophyll c* functions as an accessory pigment in brown algae, as does *chlorophyll d* in red algae. Chlorophylls absorb light energy in the red (650–680 nm) and blue (400–450 nm) portions of the spectrum (Fig. 7.2a).

Other accessory pigments absorb energy as effectively during photosynthesis as chlorophyll. Each accessory pigment requires a specific wavelength of light to be used efficiently in photosynthesis.

Carotenoids, which include the **carotenes** and **xanthophylls,** absorb in the blue-green (430–500 nm) portion of the spectrum (Fig. 7.2b) and transfer that energy to chlorophyll. They also protect chlorophyll from destruction by ultraviolet irradiation, and act as coenzymes in photosynthesis. *Beta-carotene,* which is present in brown, green, and red algae, also has major importance in brown algae (Table 7.1). *Fucoxanthin* (a xanthophyll) is a supplementary light-absorbing brown pigment in brown algae.

Phycobilins (red pigments) absorb blue, green, red, yellow (550-nm), and orange (650-nm) wavelengths (Fig. 7.2b). Red algae that live in deep water are richer in *phycoerythrin* than are shallow-water or intertidal species.

Figure 7.1 Fate of the visible portion of the electromagnetic spectrum in seawater. Absorption of light energy by seawater reduces the intensity and wavelengths available for photosynthesis by phytoplankton and attached macroalgae.

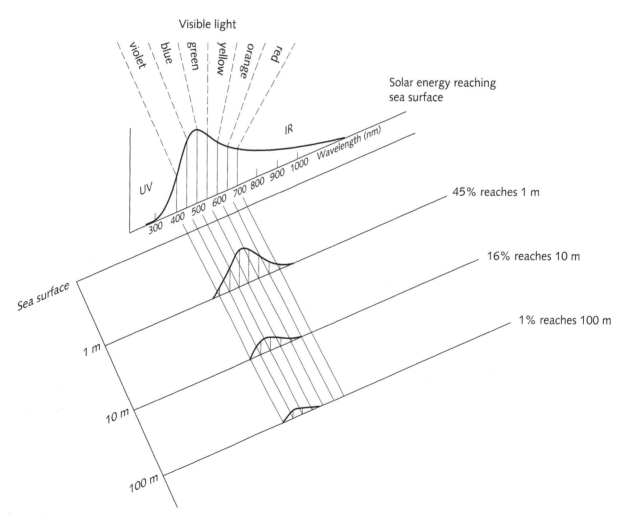

Table 7.1   Distribution of chlorophylls, phycobilins, and carotenoids in the algal divisions Phaeophyta, Chlorophyta, and Rhodophyta.

Pigment	Phaeophyta	Chlorophyta	Rhodophyta
Chlorophylls			
a	+++	+++	+++
b	−	++	−
c	+	−	−
d	−	−	+
Phycobilins			
Phycocyanin	−	−	+
Phycoerythrin	−	−	+++
Carotenoids			
Carotenes			
alpha	+	+	+
beta	+++	+++	+++
Xanthophylls			
Astaxanthin	−	+	−
Diatoxanthin	+	−	−
Diadinoxanthin	+	−	−
Flavoxanthin	+	−	−
Fucoxanthin	+++	−	−
Lutein	+	+++	++
Neofucoxanthins	−	+	−
Taraxanthin	−	−	+
Violaxanthin	+	+	−
Zeaxanthin	−	+	+

Key
- −   pigment lacking
- +   pigment present; low concentration may prevent detection by simple TLC
- ++   pigment in high concentration; usually detected by TLC
- +++ pigment in very high concentration; easily detected by TLC

a.   b.

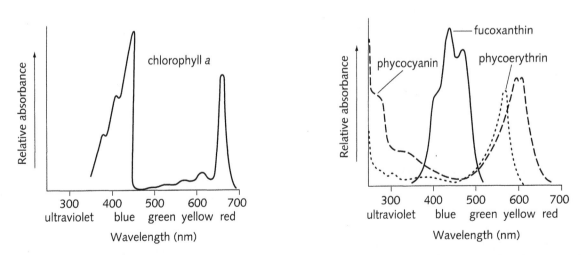

Figure 7.2   Absorption spectra of the three classes of plant pigments. (a) Chlorophyll. (b) Fucoxanthin (a carotenoid) and phycobilins (phycocyanin and phycoerythrin).

EXERCISE 1

Pigment Extraction

In this exercise you will extract fat-soluble pigments from intact pieces of macroalgae. Because the chlorophyll molecules are located in membrane-bound organelles (**chloroplasts**) within algal cells, the integrity of two membranes (organellar and cytoplasmic) must be disrupted to obtain the pigment. One common method is mechanical grinding or homogenizing, followed by centrifugation and chemical extraction. Grinding techniques are hampered by the leathery texture and mucilaginous fluids of some macroalgae. The method you will use involves hyperosmotic pretreatment of the algal sample prior to chemical extraction; it eliminates the homogenization and centrifugation steps.

Seawater of 35 o/oo salinity has a salt concentration equivalent to 0.5 M NaCl. The ionic concentrations of cell and tissue fluids of marine algae are essentially equivalent to that molarity, and are said to be **isosmotic** (= isotonic). If seawater is diluted, as normally occurs in coastal regions near estuaries, it becomes less salty (less than 35 o/oo, less than 0.5 M NaCl). It could then be **hyposmotic** compared with the cell fluids of the alga (the alga is **hyperosmotic** to the seawater). This difference in **osmotic concentration** permits water to enter the vacuole of the cell, causing it to swell.

If the ionic concentration outside the cell is hyperosmotic to the algal cell fluids, water leaves the hyposmotic vacuole of the cell. The cytoplasm of the cell condenses as the vacuole collapses, in a process called **plasmolysis.**

If hyperosmotic conditions are excessive, the integrity of the vacuolar, organellar, and cytoplasmic membranes is disrupted. Such **hyperosmotic shock** facilitates extraction of large molecules from the cell.

The hyperosmotic extraction method used here is modified from that of Reed (1988). It is applicable to the extraction of pigments from at least three specimens of algae—one each from Chlorophyta, Phaeophyta, and Rhodophyta. Each extract obtained should be sufficient for two to four students to produce absorption spectra, and to separate pigments by thin-layer chromatography.

PROCEDURE

1. Gather a set of glassware including three screw-cap vials, two 100-mL graduated cylinders, six 100-mL beakers, three Syracuse watch glasses, microscope slides, and coverslips.

2. Obtain a fresh macroalgal specimen. Blot it dry, remove a 10–50 g portion, and place it in a beaker.

3. Cover the sample with 6.0 M NaCl (= 35 g in 100 mL distilled water) for five minutes.

4. Remove the sample from the hyperosmotic NaCl solution, and blot it dry. Place it in a clean, dry beaker, and add just enough 95% methanol to cover the sample. Make sure that the sample is immersed (not floating). Stand the beaker in the dark for 30 min.

WARNING

Methanol is flammable. Keep it away from open flames or electrical outlets.

5. Examine the extract. It should be relatively dark in appearance. If it lacks sufficient color, you may need to continue the extraction with another piece of alga. Consult your instructor.

6. If you are satisfied with your extract, pour it into a screw-cap vial, and wrap it in foil. Label it, and store it in the refrigerator. Discard the algal sample.

EXERCISE 2

Absorption Spectrum of Pigment Extracts

Isolated chlorophylls and accessory pigments absorb light energy at a particular wavelength or over a range of wavelengths. These peaks in the **absorption spectrum** tend to be in the blue-green (400–500 nm) and far-red wavelengths (600–700 nm) (Figs. 7.2 and 7.3).

In this exercise, you will attempt to produce an absorption spectrum for each of your pigment extract samples. You will use a **spectrophotometer,** an instrument that detects the amount of light that is absorbed by the algal extract at different wavelengths.

PROCEDURE

1. You will be instructed how to use a spectrophotometer. Do not attempt to use it until you have been briefed on its operation.

2. Obtain a set of **cuvettes** (special photometer tubes for use in the spectrophotometer). With a glass marking pencil, label one **B** for **blank** near

a.

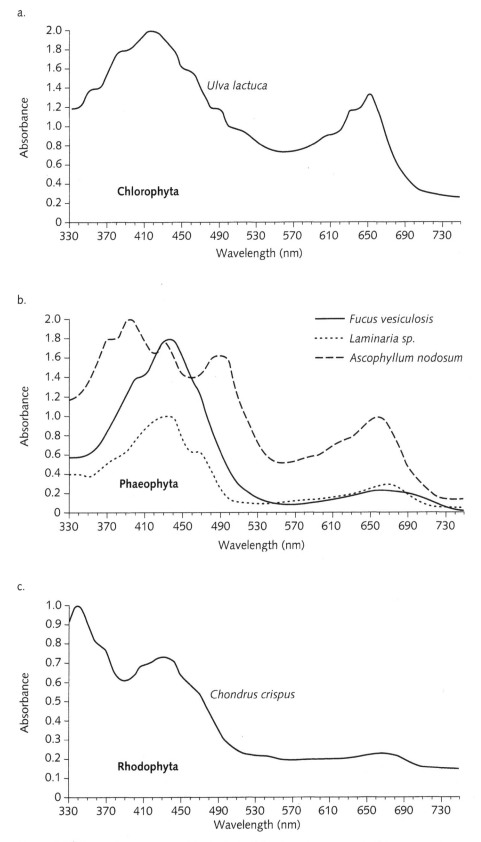

Figure 7.3 Absorption spectra of five species of macroalgae representing Chlorophyta (a), Phaeophyta (b), and Rhodophyta (c). All extracts in 95% methanol following hyperosmotic pretreatment.

the rim of the tube. Label the others in a way that designates the different pigment extracts that you will be reading.

3. Add sufficient volume of 95% methanol to the **B** cuvette to bring the fluid level up to the horizontal mark on the cuvette, or to two-thirds of the capacity of the cuvette.

4. Add equivalent volumes of your pigment extracts to the other labeled cuvettes.
NOTE: If your extracts are too dark, they will absorb too much light and fail to give you a readable absorption spectrum. You may need to dilute your extracts with 95% methanol prior to reading. Consult your instructor.

5. Adjust the wavelength selector of the spectrophotometer to 350 nm.
   a. Insert the blank cuvette in the instrument and adjust the absorbance reading to zero (note that zero absorbance = 100% transmittance).
   b. Remove the blank, set it aside, and insert a sample cuvette. Read the absorbance value, and record it on Report Form A.

6. Repeat the procedures described in step 5 at 25-nm wavelength intervals from 350 nm to 750 nm. For each new wavelength, insert the blank cuvette, and adjust the reading to zero prior to reading the sample absorbance.

7. Dispose of your extracts and solvents in the waste container designated for chemical waste.

8. Use the recorded absorbance readings and plot the absorption spectrum on the graph paper on Report Form A.

Interpretation

1. What color are your extracts? Explain why you see those colors.

2. Examine the plot of your absorbance spectra. Do you detect any absorbance peaks in your spectra? If so, indicate the ranges of wavelength and colors that are most effectively absorbed by your extracts.

3. Examine Figure 7.2. Which of the three major categories of plant pigments could be represented by your absorption spectra? Explain.

4. Compare your spectra with those illustrated in Figures 7.2 and 7.3. Can you explain any observed differences?

5. Compare the spectral quality of solar radiation in the Gulf Stream (Fig. 7.4) with absorption spectra of your samples. Try to explain any similarities or differences in the patterns.

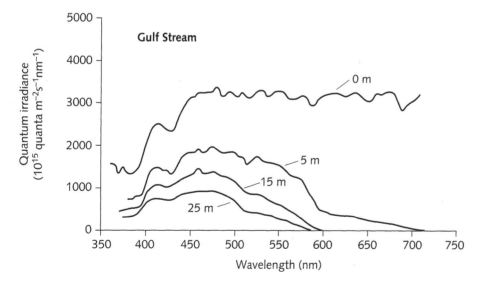

Figure 7.4 Spectral quality of solar radiation at four depths in the Gulf Stream. From Adey and Loveland, 1991.

EXERCISE 3
Separation of Plant Pigments

According to Table 7.1, several different pigments can be present in any one species of alga. Multiple pigments may be difficult to discern in intact specimens because the colors of some pigments are masked by the colors of others. Even the extracts appear to be uniformly green.

It is possible to demonstrate the existence of several pigments in your extract by using the technique of **thin-layer chromatography (TLC).** In this case, the *thin layer* is a coating of Silica Gel G that is placed on a glass slide or acetate sheet. Chromatography refers to the measurement of the relative distances between pigment bands that appear on the coating during the procedure.

In this procedure, the pigment extract (prepared earlier) is applied to the gel. The pigment molecules are adsorbed to the gel particles, and the methanol solvent evaporates. The gel slide or sheet is placed in a **chromatographic chamber** containing a solvent that travels across the silica gel surface. As the solvent passes through the adsorbed pigment molecules, the molecules dissolve in the solvent and are carried along with it. Pigment molecules that dissolve quickly in the solvent or that have relatively low molecular weights move rapidly through the gel with the solvent. Larger molecules or those that dissolve slowly in the solvent move less rapidly. This differential transport and movement separates the pigments from one another and allows them to be identified by their relative final positions on the TLC strip.

Under a given set of conditions (including type of solvent), each pigment will usually travel at the same rate, known as the **rate of flow (Rf)**. The Rf is a ratio of the distance traveled by the pigment (D_p) to the distance traveled by the solvent (D_s), or:

Rf = D_p / D_s

Because each pigment travels at a different rate, each pigment can be identified and distinguished from all other pigments by its unique Rf (Tables 7.2 and 7.3). Even different pigments with the same apparent color can be distinguished by their different Rf values.

In this exercise you will use TLC to separate and identify pigments contained in your algal extract.

PROCEDURE

1. Obtain silica gel strips, capillary tubes, and a Coplin jar (Fig. 7.5), which will serve as the chromatographic chamber.

Table 7.2 Thin-layer chromatography of *Ulva lactuca* (Chlorophyta) methanol extracts in two solvent systems.*

Rf	Pigment Color	Pigment Identification**
Solvent System 1		
0.03	yellow	xanthophyll
0.10	yellow	neoxanthin
0.13	yellow	chlorophyll *b*
0.17	yellow	xanthophyll
0.25	blue-green	chlorophyll *a*
0.92	yellow	beta carotene
Solvent System 2		
0.05	yellow	bixin
0.56	gray	ethyl 8'-apo-beta carotenoate
0.96	yellow	lycopene or beta carotene

*Solvent System 1 (84% petroleum ether, 15% acetone, 1% isopropanol); Solvent System 2 (80% petroleum ether, 20% benzene, 2% acetone, 1% acetic acid).
**Identification based on Kirchner (1978).

Table 7.3 Thin-layer chromatography of *Fucus vesiculosus* and *Ascophyllum nodosum* (Phaeophyta) methanol extracts in solvent system 2.*

Rf	Pigment Color	Pigment Identification**
Fucus vesiculosus		
0.03	yellow	capaxanthin
0.08	yellow	bixin
0.12	yellow	crocetin
0.38	green-yellow	8'-apo-beta carotenal
0.46	green-yellow	methyl 8'-apo-beta carotenoate
0.96	yellow	lycopene or beta carotene
Ascophyllum nodosum		
0.08	yellow	bixin
0.11	yellow	canthaxanthin or capaxanthin
0.57	gray-yellow	ethyl 8'-apo-beta carotenoate
0.97	yellow	lycopene or beta carotene

*Solvent System 2 (80% petroleum ether, 20% benzene, 2% acetone, 1% acetic acid).
**Identification based on Kirchner (1978).

Figure 7.5 Coplin staining jar as chromatographic chamber holds regular 75 mm × 25 mm silica gel strips.

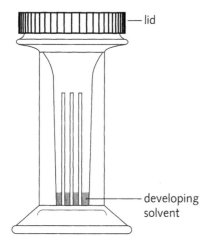

lid

developing solvent

Figure 7.6 Silica gel slide (strip). (a) Strip with spotted pigment extract line (origin is 1 cm from one end). (b) Developed strip illustrating origin, solvent front, and several pigment bands (with accompanying Rf values).

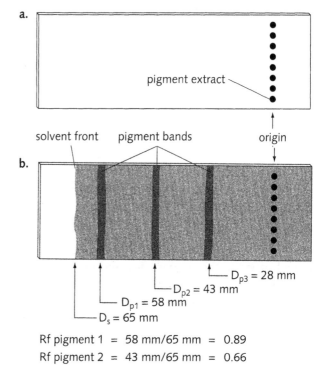

a.

pigment extract

solvent front   pigment bands   origin

b.

D_{p3} = 28 mm
D_{p2} = 43 mm
D_{p1} = 58 mm
D_s = 65 mm

Rf pigment 1 = 58 mm/65 mm = 0.89
Rf pigment 2 = 43 mm/65 mm = 0.66
Rf pigment 3 = 28 mm/65 mm = 0.43

2. With a pencil, make a small tick-mark on the edge of the gel surface of the strip 1 cm from one end. This mark indicates the location where you will apply your extract.

3. Dip one end of the capillary tube in your extract and allow the tube to fill.

4. Apply the extract to the gel strip.

   a. Hold the capillary tube at a slight angle and gently cover the upper end of the tube with your index finger.

   b. Lightly touch the lower end of the tube to the gel surface near the location of your pencil mark, but not at the edge (Fig. 7.6a). A very small quantity of extract should flow onto the gel.
   NOTE: Each drop should not be more than 2 mm in diameter. *Do not scratch the surface of the gel.*

   c. Apply 10–15 drops adjacent to one another to form a line across the strip (Fig. 7.6a).

   d. Allow the drops to dry, then repeat the process two or three times.

5. Develop the **chromatogram.**

   a. Place about 2 mL of developing solvent* into the Coplin jar (Fig. 7.5). The level of the solvent must be *below* the line of pigment spots on the TLC strip.

   WARNING

   Developing solvents are flammable. Keep them away from open flames and electrical outlets. *Do not inhale the fumes.*

   b. Cover the jar for one or two minutes. Use the screw cap, if available. If you are using Coplin jars with ground-glass covers, place a layer of Parafilm over the top of the jar to ensure a tight seal, and then secure the glass cover.

*You may be asked to use two different solvent systems. If so, duplicate your apparatus.
   Solvent System 1 separates chlorophylls: petroleum ether, acetone, isopropanol, in volume proportions (84:15:1).
   Solvent System 2 separates carotenoids, but is not effective with chlorophylls: petroleum ether, benzene, acetone, acetic acid, in volume proportions (80:20:2:1).

c. Place the gel strip, pigment line down, into the Coplin jar. Cover the jar again.
   NOTE: Do not insert the edges of the strip in the glass grooves in the jar—doing so will distort the progress of the solvent front.

d. Observe the chromatogram frequently. When the solvent front (Fig. 7.6b) moves to within 0.5 cm of the top of the gel strip (about 5 min), remove the strip, and cover the jar.

e. *Immediately* make a pencil mark on the strip to identify the position of the solvent front.

f. Measure (in mm) D_s (distance between the original pigment line and the solvent front). Record this value on Report Form B.

g. Measure (in mm) D_p for each newly separated pigment band (distance from the original pigment line to the center of each pigment band). Record these values, and the corresponding pigment color on Report Form B.

h. Calculate the Rf for each pigment band, and record these values on Report Form B.

Interpretation

1. Several developing solvent systems are available; each gives different results in terms of pigment solubility and separation.

   a. Compare your chromatograms to the information in Tables 7.2 and 7.3. You may not see all of the bands listed, or you may notice some unidentified pigments.

   b. Try to identify the pigment bands on your chromatograms.

REFERENCES

Adey, W. H., and K. Loveland. 1991. *Dynamic aquaria.* Academic Press, San Diego, California.

Kirchner, J. G. 1978. *Thin-layer chromatography.* Wiley and Sons, New York.

Reed, R. H. 1988. Hyperosmotic pretreatment of marine macroalgae prior to extraction of chlorophyll in methanol and dimethylformamide. *Phycologia,* 27:477-484.

REPORT FORM A

Division _____ Species _____

EXTRACT:

Weight of Sample _____g   Volume of Methanol _____mL

Color of Extract _____

ABSORPTION SPECTRUM

COLOR	WAVELENGTH (nm)	ABSORBANCE
UV	350	————
	375	————
Violet	400	————
	425	————
Blue	450	————
	475	————
Yellow	500	————
	525	————
	550	————
	575	————
Orange	600	————
	625	————
	650	————
Red	675	————
	700	————
	725	————
	750	————

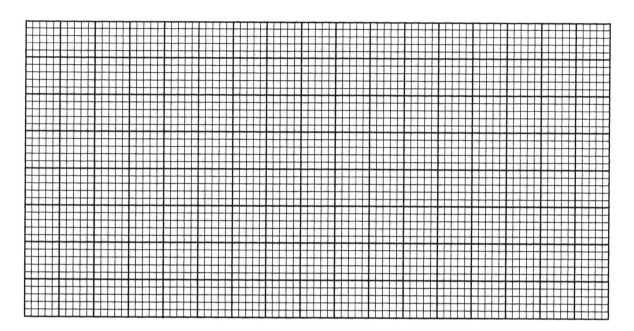

Legend of Graph:

REPORT FORM B

Division ——————————————————— Species ———————————————————

PIGMENT SEPARATION

Solvent System ————————————————————————————————————

D_s ———————————————————————————————— mm

Pigments
——

D_p (mm)	Rf	Color	Identification
	0.		

Plankton

OBJECTIVES

After completing this unit, you will be able to

- Recognize the basic characteristics of representative phytoplankton (diatoms and dinoflagellates), herbivorous zooplankton (copepods), and predator zooplankton (arrow worms);

- Appreciate the diversity of marine plankton by observation of preserved samples;

- Know how to identify plankton with the aid of simplified keys and illustrations;

- Estimate relative abundance of plankton in a sample;

- Recognize the diversity of individuals in a living marine plankton sample and observe locomotion and behavior of live plankters;

- Determine the relative abundance of dominant species within the community sampled; and

- Estimate the biomass (standing crop) of a plankton community by analysis of a field sample.

INTRODUCTION

Plankton are organisms that live freely in seawater and that are unable to swim against a current of one knot (nautical mile per hour) for an extended period. Because planktonic organisms are quite diverse in size, shape, and biological origin, they can be categorized in several ways:

Basic Nutritional Requirements

Phytoplankton—contain chlorophyll and are photosynthetic. Six plant divisions are represented: Cyanophyta or Cyanobacteria (blue-greens), Chlorophyta (green algae), Phaeophyta (brown algae), Rhodophyta (red algae), Chrysophyta (golden-brown algae, or diatoms and coccolithophores), and Pyrrophyta (dinoflagellates). The latter two divisions are the most commonly encountered groups of marine phytoplankton.

Zooplankton—heterotrophic, animal plankton. Virtually all marine animal phyla are represented in the zooplankton, either throughout their life cycles or as larval stages (Fig. 8.1).

Length of Planktonic Life

Holoplanktonic—planktonic at all stages of the life cycle (for example, copepod) (Fig. 8.1).

Meroplanktonic—species that are planktonic for only a portion of the life cycle (for example, barnacle) (Fig. 8.1).

Tycoplanktonic—small animals, principally benthic, that are suspended temporarily into the water column by currents, behavior, or other mechanisms.

Habitat

Oceanic—plankton occurring in the oceanic province (the mass of water beyond the influence of the continental shelves).

Neritic—plankton in waters over the continental shelves.

Epipelagic—plankton occurring in surface waters (that is, to depths of 200 m).

Mesopelagic—plankton common at depths from 200 m to 1000 m.

Neuston—plankton living in or near the surface film of the water.

Size

The samples that you will examine in this unit will vary according to the mesh size of the collecting net

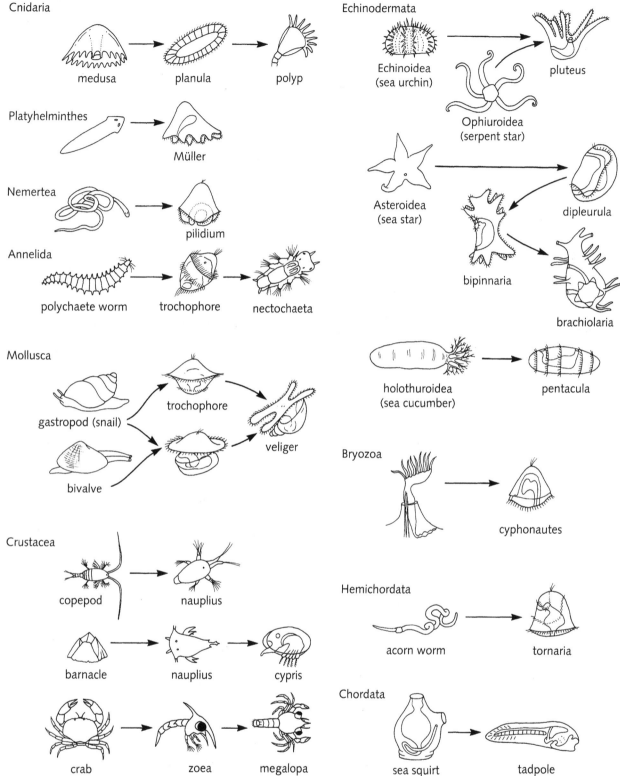

Figure 8.1 Planktonic larval stages in marine invertebrate life cycles. Except for the pelagic cnidarian medusa and the planktonic copepod, all adults (on the left in the cycles) are benthic organisms.

(see Table 8.1 and Fig. 8.2). For example, if you use a No. 80 Nitex net, you will encounter a large array of larger microplanktonic phytoplankton as well as macro-zooplankton. Samples collected with a coarser net, such as No. 571 (mesh size 0.571 mm), will contain relatively large zooplankton and few phytoplankton. The following prefixes categorize size:

Mega-	larger than 2 mm
Macro-	larger than 0.5 mm
Meso-	0.2–0.5 mm
Micro-	0.06–0.202 mm
Nano-	0.005–0.06 mm
Ultra-	less than 0.005 mm

The exercises in this unit are designed to acquaint you with plankton from several perspectives. You can
1. Study examples of autotrophic, herbivorous, and carnivorous plankton;
2. Observe the diversity of form commonly present in plankton collections;
3. Observe the fascinating world of living plankton; and
4. Make a quantitative assessment of a sample.

It is unrealistic to expect to accomplish all of these goals in one or two laboratory periods. Your instructor will inform you which exercises will be performed in your course.

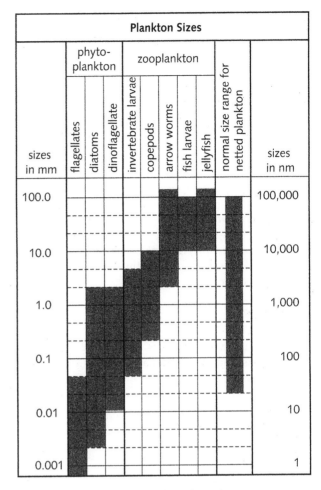

Figure 8.2 Range of sizes for selected taxonomic groups of plankton. Note that size scales are logarithmic.

TABLE 8.1 Standard sizes of plankton nets made of bolting cloth or Nitex and their mesh dimensions.

| Silk Bolting Cloth Size | Nitex Nylon Screen Cloth | Mesh Measurements | | | |
		Meshes per inch	% Open Area	μm	Millimeters (mm)
—	4000	5	64%	4000	4.000
0000	1340	16	59	1340	1.340
000	1050	23	59	1050	1.050
0	571	38	50	571	0.571
2	263	54	47	263	0.263
6	243	74	45	243	0.243
10	153	109	45	153	0.153
20	80	198	38	80	0.080
25	64	227	32	64	0.064
—	30	390	22	30	0.030
—	10	420	5	10	0.010

A. Representatives of the Plankton Community

The plankton community serves as an important source of food for a variety of marine organisms, from benthos to nekton. Often overlooked is the existence of food webs within the plankton community itself.

EXERCISE 1

Phytoplankton

The autotrophic phytoplankton form the basis of the food chains that extend through the herbivorous zooplankton to the predators.

PROCEDURE

1. Examine living cultures or prepared slides of diatoms and dinoflagellates with a compound microscope.

a. **Diatoms** (Fig. 8.3) are enclosed within a unique "pillbox" constructed of silicon dioxide (glass) and have no visible means of locomotion. Each box is highly ornamented with species-specific pits and perforations. The diatoms may occur singly (each individual to a box) or in chains of various kinds.

b. **Dinoflagellates** (Fig. 8.4) have two flagella—one is wrapped around a groove (girdle) along the middle of the cell, the other trails free. These organisms lack an external skeleton of silicon, but are armored with plates of cellulose. Dinoflagellates are usually small and solitary.

2. Sketch several specimens, and measure two dimensions with the ocular micrometer (refer to Appendix I for directions on the use and calibration of the micrometer).

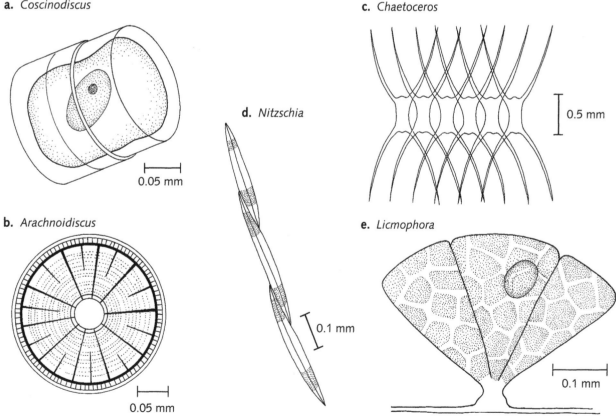

a. *Coscinodiscus*

0.05 mm

b. *Arachnoidiscus*

0.05 mm

d. *Nitzschia*

0.1 mm

c. *Chaetoceros*

0.5 mm

e. *Licmophora*

0.1 mm

Figure 8.3 Diatoms (Chrysophyta). (a, b) Centric diatoms with cylindrical or drum-shaped frustule. (c) Some centric diatoms form chains. (d, e) Pennate diatoms with tapered or triangular frustules.

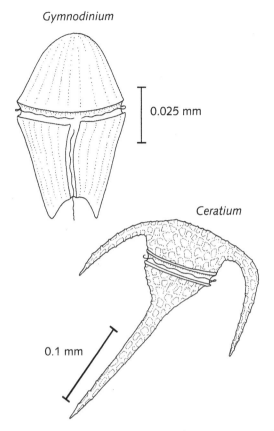

Gymnodinium

0.025 mm

Ceratium

0.1 mm

Figure 8.4 Dinoflagellates (Pyrrophyta) are unusually shaped organisms with two flagella lying in conspicuous depression (girdle).

E X E R C I S E   2

Grazers: The Copepods

Copepods (oar-footed) are small crustaceans (from less than 1 mm to several millimeters in length) that lack both compound eyes and a carapace. Although copepods have a varied morphology that corresponds to different modes of life, most species are free-living, motile zooplankters that feed extensively on phytoplankton. This largest and most important group of meso-zooplankton serves as a vital link between phytoplankton and larger carnivores. **Calanoid** copepods (Fig. 8.5) dominate oceanic systems with the greatest absolute abundance and biomass.

PROCEDURE

1. Obtain samples of copepods. Samples may include prepared slides (whole mounts), living organisms, and preserved plankton collections.

   a. Place entire *preserved* specimens in a depression slide or a small petri dish.

   b. If you are using *living* specimens, place them in the well of a depression slide, or prepare a Vaseline wet-mount (see Unit 4).

2. Study your specimens as you read the descriptions that follow. Consult the illustrations in this unit and available reference texts as you work. Record your observations in your notebook.

3. Begin your observations with a dissecting microscope at low magnification. For more detailed observations, use the compound microscope.

4. If your microscope is equipped with an ocular micrometer, measure the length of the specimen. Refer to Appendix I on the use of microscopes for information on the calibration and use of micrometers.

5. Draw your specimen as you see it. Do not assume that it will resemble the one illustrated in this manual.

6. With fine-tipped forceps, try to remove the mouthparts from one side of the body and mount them in water (substitute glycerin or mineral oil if necessary) on slides for observation at higher magnification with a compound microscope.
   NOTE: If you have trouble with your surgery, make a "squash" preparation of the animal. Place the specimen in a few drops of water (or glycerin) on a coverslip. Place a microscope slide on top of the animal, and gently press and move the slide a little to the side. This should expose the appendages for better viewing. If necessary, remove the coverslip and add more fluid. Using dissecting needles, separate the appendages from each other and from the body.

7. Draw a mouthpart observed under high power. Emphasize setal detail in your drawing. Can you distinguish between the larger **spines** and smaller **setae?** Are **setules** (small setae) present? Label the drawing as fully as you can.

8. With an ocular micrometer, measure the length of the mouthpart, and the distance between two neighboring spines or major setae on one of them. How do these measurements correspond with the size of diatoms and dinoflagellates?

DESCRIPTION

The typical copepod contains 10 **trunk segments (somites)** and a terminal segment **(telson)** with two **caudal rami** (Fig. 8.5). The body is divided into an anterior appendage-bearing part **(prosome)** and a posterior part **(urosome).** The anterior projection of the head, called the **rostrum,** may project forward as a short beak or may be slung under the head. If a simple **naupliar eye** is present, it is medially located at the anterior-most portion of the head.

The **first antennae** are unbranched **(uniramous)** and often very long. They act as stabilizers of body position. The first antennae of males are modified into a grasping structure for holding females during copulation. The **second antennae** are shorter than the first, and typically forked **(biramous).** They produce water currents during feeding and slow swimming.

The rodlike and segmented urosome lacks appendages, but has two caudal rami (singular = ramus) at its extremity. The rami act as a steering vane during rapid escape movements.

The four pairs of mouthparts have a diverse morphology. Some are biramous and may be heavily setated. The **setae** of filtering appendages usually contain finer processes, called **setules.** Some have stout **spines** that are used to clean the filtering setae of other mouthparts.

The first five thoracic limbs—called **swimming legs**—are biramous, flattened, and paddlelike. These paired limbs are often joined by a **medial plate** that causes both legs in a pair to move as a functional unit. The last pair often differ between the sexes: In females, they may be absent or reduced; in males, they are modified, and are used to transfer spermatophores during copulation.

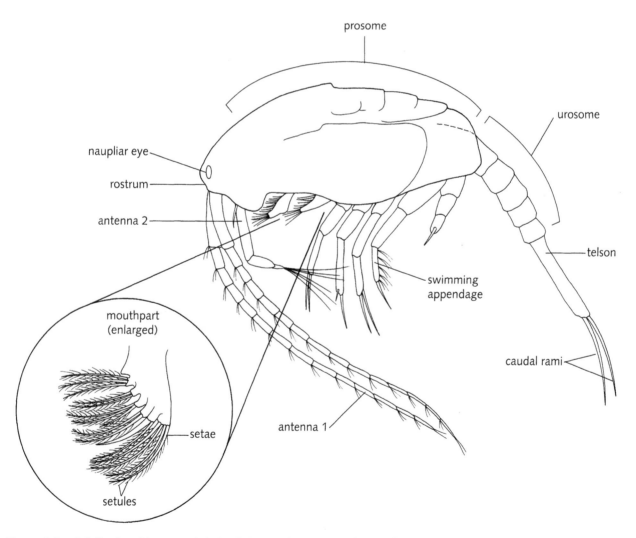

Figure 8.5 Adult calanoid copepod. Lateral view with swimming legs in the anterior position. Enlarged image of one of the filtering mouthparts shown at left.

EXERCISE 3

Predators: Arrow Worms

Members of the phylum Chaetognatha (bristle jaw) are commonly called arrow worms because of their shapes. Some 70 species, all marine, can be commonly found in tropical and subtropical waters. Most are planktonic and effective predators.

PROCEDURE

1. Examine preserved specimens or prepared slides, and read the following description, which is based upon *Sagitta elegans*, a very common species.

2. Draw and measure the specimen.
   a. Measure the length.
   b. Measure the distance between the left and right sets of grasping spines (see below). What kinds of plankton could be seized by your specimen?

The complex **head** region is equipped with feeding and sensory organs. If a **hood** covers the head (Fig. 8.6), try to remove it with dissecting needle and forceps. See if you can locate the pair of **eyes,** which are dorsal and medial in position.

Locate the large chitinous **spines** used to seize prey. They are operated by muscles attached to plates in the head. **Jaws** armed with **teeth** grasp food, which is then pushed into the mouth and pharynx. The intestine is a simple tube that opens posteriorly near the caudal fin.

Find the two pairs of lateral fins and the caudal fin. The fins have raylike supports. With proper illumination of cleared specimens, you may be able to see longitudinal muscle bands running the length of the animal.

Interpretation of Representative Plankton

1. Copepods are setular filter feeders. What does that mean?

2. For the copepod, correlate the spaces between setae and setules on the feeding appendages with the size and type of food that potentially can be captured. Consult Table 8.1 and Figure 8.2, and consider the size of the phytoplankton that you observed.

3. How effective a predator is the arrow worm? How difficult (or easy) would it be for the arrow worm to consume the copepods you observed? Can you support your answer with data on measurements?

4. What organisms eat chaetognaths?

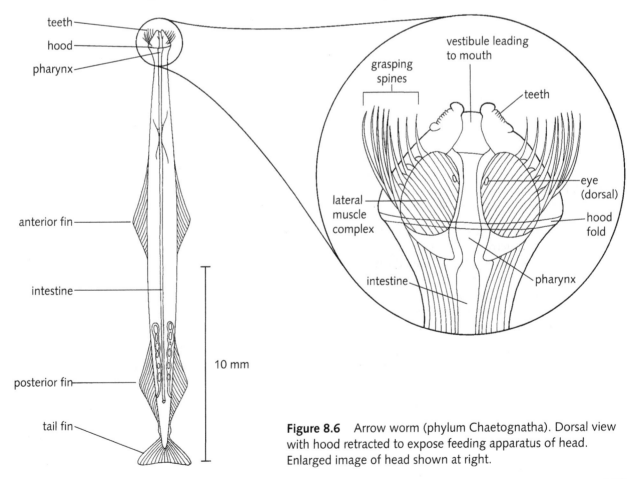

Figure 8.6 Arrow worm (phylum Chaetognatha). Dorsal view with hood retracted to expose feeding apparatus of head. Enlarged image of head shown at right.

B. Composition of the Plankton

Identification of marine plankton is both challenging and fascinating. Variation in form and the myriad changes that occur from the egg through larval forms to the adult make it difficult to devise a detailed systematic key. In most research laboratories engaged in studies of plankton, scientists use illustrated monographs and plates for identification. Few marine biologists are familiar with the broad spectrum of planktonic species; instead, most specialize in a particular group of plankton.

Although a plankton community can be diverse in terms of the number of species present, only a few taxonomic groups dominate the holoplankton. The most important groups of phytoplankton are diatoms and dinoflagellates. Copepods, krill, and other crustaceans dominate the zooplankton. Copepods are the main herbivores, and by far the most abundant group of zooplankton. Many groups of animals other than crustaceans—most of them carnivorous—are present in the zooplankton. Arrow worms (chaetognaths) are extremely important predators. Salps and larvaceans are filter feeders.

Meroplankton provide a contrast. A vast number of organisms have planktonic eggs and larvae. Coastal waters are especially rich in meroplankton, particularly during peak reproductive activity.

EXERCISE 1

Identification of Marine Plankton

PROCEDURE

1. Examine samples of plankton netting under the dissecting microscope. Pay close attention to the fabric of the net. Is it constructed of silk, nylon, or another polymer? Is it woven?

   a. Measure the mesh size of net samples with the ocular micrometer of the compound microscope. If necessary, calibrate the micrometer (refer to Appendix I on the use of microscopes). Record your measurements at the top of the Composition of Plankton Report Form.

   b. Compare the measurements of your net samples with the dimensions listed in Table 8.1. Consult Figure 8.2, and list (on the Report Form) those groups of plankton most likely to be collected with nets of those mesh sizes.

2. Obtain a binocular dissecting microscope outfitted with a glass stage plate and a substage mirror. If you are unfamiliar with this instrument, read the directions for its use before proceeding (see Appendix I). Mount the lamp in the base of the microscope, and adjust the mirror to achieve maximum contrast of plankton against the background. Begin your observations at the lowest magnification possible.

3. With a large-bore pipet or medicine dropper, dispense 1 mL to 2 mL of a plankton sample into a small (35 mm diameter) petri dish or a Syracuse dish.

CAUTION

Plankton samples are routinely preserved in a 5% solution of neutral formalin. This solution is an irritant. Care must be taken not to spill it or get it in your eyes. Wipe up spills and wash your hands when you are finished with the exercise.

An alternative approach is to retain the collection in a sieve, and wash it with filtered tapwater. Formalin can be added at the end of the exercise. The sieving must be done carefully to prevent damage to the plankters.

4. Add sufficient **neutral formalin** to cover the bottom of the dish. Gently agitate the dish to distribute the plankton thinly and evenly over the bottom of the dish.

5. Start in the part of the dish farthest from you. Use the illustrated Keys to Plankton provided with this unit (see page 126) and assign the first organism to one of the taxonomic groups designated in the Keys. After the instructor confirms your identification, proceed with the exercise.

   a. Determine if the specimen is an adult or a larval stage.

   b. Determine if the specimen is holoplanktonic (planktonic at all stages of life cycle) or meroplanktonic (planktonic as egg, larva, or juvenile).

c. Scan the entire sample and look for similar specimens. As you proceed, estimate the relative numerical abundance. Use the following scale:

present	= +	= from 1 to 10
common	= + +	= more than 10; less than 50
numerous	= + + +	= more than 50; less than 100
abundant	= + + + +	= more than 100 (too numerous to count)

NOTE: A more accurate quantitative assessment is described in Part D of this unit, "Quantitative Analysis of a Plankton Sample."

6. Tally your observations in the Plankton Checklist (Table 8.2 in Part D of this unit), or in your notebook.

7. Repeat steps 5 and 6 for each taxon.

8. Return the organisms to the bottle from which they were obtained. Flush them from the dish with neutral formalin dispensed from designated wash bottles or medicine droppers.

9. Dispose of any waste formalin in the designated container. *Do not flush it down the drain!*

10. **Wash** your glassware with detergent and rinse it well.

C. Observation of Living Plankton

Observing living plankton is fascinating. The thrill of seeing the color and activity of the organisms outweighs the slight frustration of attempting to focus on a highly motile specimen. Your patience will be challenged during the next exercise, but you will be rewarded by the visual stimulation that you receive.

Interpretation

1. Construct a food web of the planktonic ecosystem represented by your sample. Place each taxonomic group that you identified in its correct feeding category (autotroph, herbivore, carnivore). Diagram as many feeding levels as you think exist in your sample. Using arrows, draw feeding pathways from one group to another, showing the various predator–prey relationships. Arrows point to the consumer.

2. Add any other nonplanktonic organisms that you think might be involved in the food web.

3. Indicate nonorganic nutrients necessary for primary producers, and show how the nutrients return to the system by decomposition of the trophic levels.

4. This exercise did not emphasize a quantitative analysis of a plankton sample. Assume that your sample accurately represented the plankton community at the time of sampling.

   a. Which groups were dominant?

   b. Which predominated, holoplankton or meroplankton?

   c. Which predominated, adults or eggs/larvae?

   d. Can you establish any correlations of your answers to a, b, and c with mesh size of collecting net, or with location and time of sampling?

EXERCISE 1

Observing Diversity, Locomotion, and Behavior of Plankton

PROCEDURE

1. Work independently. Procedures are similar to those used in the identification of preserved samples, except that you will be working with live organisms in seawater, and you *must use clean glassware.*

2. Obtain a binocular dissecting microscope, a compound microscope, petri dishes, Syracuse dishes, depression slides, regular slides, and coverslips.

3. Place a subsample of plankton in a petri dish (in sufficient seawater to cover the bottom). Concentrate the stock sample to ease retrieval or capture of the living "plankters."

   a. Direct a beam of light into the dish from the side. Positively phototactic organisms will congregate near the light source; negatively phototactic ones will move to the opposite side of the container.

   b. Use a pipet or medicine dropper with a wide opening to capture the plankters and transfer them to the Syracuse dish.

4. Begin your observations with the dissecting microscope, then switch to the compound microscope when higher magnification becomes necessary.

   a. If plankters move too fast for viewing in the Syracuse dish, concentrate them with a light source, and transfer them to a depression slide or a Vaseline wet-mount (see Unit 4 for directions).

   b. Add a drop or two of methyl cellulose or a commercial preparation (Protoslo or Detain) to the slide to retard movement.

5. *Observe, describe, and illustrate.* No set procedure exists for this part of the exercise, but your laboratory report should include a description of the specimens you observe, their features, behavior, and movement. You might select one particular aspect and make a comparison among several different species. Try to associate what you see with some adaptive advantage of the trait. Consider the following characteristics as you work:

   a. Color. Describe the pattern. Is it uniform? Banded? Striped? Is the color confined to chromatophores (pigment cells)? Are there different colored chromatophores? How are they distributed on the body?

   b. Describe the shape of the specimen. Does it have appendages? How many? Where are they located? What purpose do they serve? Do they have setae?

   c. Describe the activity and locomotion. Is the movement jerky and sporadic? Does the animal glide? Are ciliated bands present?

   d. Is there any indication of feeding? How is it accomplished? What is the food?

   e. Are eyes present? How many? Where? Do you see any evidence of phototropism?

Interpretation

1. Describe, in writing, your observations of living plankton. Illustrations are encouraged and should be added to your report.

D. Quantitative Analysis of a Plankton Sample

In any study of a plankton community, the biomass and relative abundance of different species within the community must be assessed. If you know the volume of water through which the net is towed, and you determine the biomass or standing crop of the plankton in the collection, you can calculate the density of the plankton community. The density of plankton in a given volume of seawater provides a measure of their availability as food for other organisms.

Certain assumptions—some valid, others invalid—are inherent to plankton analyses. The assumptions include the following:

- Individuals in the plankton are distributed evenly in the water mass sampled;
- Collection methods do not disrupt that distribution;
- The collecting net samples all species equally;
- The net does not selectively exclude organisms;
- The sample collected accurately represents the natural community;
- Laboratory manipulations (splitting, subsampling) yield reliable subsamples (aliquots) of the natural community; and
- All identifications and counts are error-free.

In spite of the apparent unrealistic nature of these assumptions, consistent, repeated sampling does provide a reasonably reliable measure of the change in the community over time. This exercise introduces this aspect of planktonology.

EXERCISE 1

Determining Biomass and Relative Abundance within a Subsample

PROCEDURE

1. Record data pertinent to the plankton *collection** in Section A of the Plankton Analysis Report Form. The instructor will provide you with these data if they are not included on the accompanying label.

2. Measure the volume of the preserved plankton collection.

   a. Pour the collection into a graduated cylinder that has a capacity larger than that of the sample.

   b. Rinse all organisms from the collection jar into the cylinder. Use the same solution as that used for the sample preservative.

CAUTION

Plankton samples are routinely preserved in a 5% solution of neutral formalin, an irritant. Take precautions not to spill it or get it in your eyes. Wipe up spills immediately and wash your hands after handling the material.

   c. Allow the collection to settle for at least 15 min and then measure (and record) the **total volume of the collection** (Line B in Section B on the Plankton Analysis Report Form), and the **volume of the settled plankton** (Line C on the Report Form).

   d. Return the entire sample to the original collection jar.
   NOTE: If the **total volume** of the sample exceeds that of the collection jar, allow the plankton to settle in the jar, and pour the excess fluid into another container. *Do not pour formalin down the drain!*

*In this exercise, the terms *"collection"* and *"sample"* refer to the plankton collected and preserved in the field. The term *"subsample"* refers to some small portion removed from the collection or sample.

3. Subsample the collection so that you reduce the number of individuals in the sample to approximately 100–150 individuals. Two methods for reducing the sample are described here.

   a. Use a **Folsom plankton splitter** (Fig. 8.7a). This device (standard size 0.5 L capacity) enables you to split the collection literally in half. Each half can be split repeatedly until the final subsample has sufficiently few individuals for ease in processing in the allotted time.

      1. Place the collection into the undivided portion of the rotating splitter drum.
      2. Rotate the drum until all of the sample is poured into the two removable trays.
      3. If you need to subdivide the sample further, return the contents of one of the trays to the splitter drum. Repeat the preceding step.
      NOTE: You may need to add more preservative to the sample as subdividing is repeated.
      4. Record the number of times you split the sample on Line E of Section B on the Plankton Analysis Report Form. Your final computations and estimate of plankton abundance depend upon this number. For example, if you split the sample one time, you halved the sample. Your counts must be multiplied by 2. If you split the sample twice, your counts are made on ¼ of the original sample (½ of ½). Your counts must be multiplied by 4.
      5. Go to Step 4 of this exercise.

   b. Subsample with a **Hensen-Stempel pipet** (Fig. 8.7b), which enables you to remove a known volume (and hence, a known proportion) from the collection. Several sizes of pipets and spool inserts are available for use: 1, 2, 5, 10, 25, 50 mL. Your instructor will indicate the volume that is compatible with the sorting tray or counting chamber that you will be using.

      1. Record the **spool volume** of the insert that you use (Line E in Section B on the Plankton Analysis Report Form).
      2. Open the Hensen-Stempel pipet by pushing the plunger all the way down; this forces the insert out of the barrel. Agitate the sample with the pipet to achieve a well-mixed slurry, but *do not stir or swirl it.* Such actions have sorting actions of their own.

3. While the sample remains suspended, place the opened pipet into the center of the slurry and close the pipet by pulling the plunger (and insert) into the barrel.
4. Remove the pipet and rinse off any organisms adhering to the outside.
5. Go to Step 4 of this exercise.

4. Discharge the subsample into a counting chamber or a small (35-mm diameter) petri dish with a grid on the bottom.

5. Examine your subsample with a binocular dissecting microscope.

   a. Add sufficient 5% neutral formalin to distribute your sample as evenly as possible throughout the counting chamber.
   NOTE: When rinsing subsamples from container to container, use the wash bottles containing *5% neutral formalin only.* Do not dilute the collection with water!

6. Identify the plankton in your subsample.

   a. Place the counting chamber on the stage of a binocular dissecting microscope. Begin your observations in the upper left grid and work your way to the right, one grid at a time. When you reach the far right grid, move to the next line and scan the grids from right to left. Continue this sequence until you have reached the bottom, and last, grid.
   NOTE: If the chamber you are using has no grid, score the bottom of a plastic petri dish with a dissecting needle.

   If you are using a counting wheel, place it on the stage of the microscope so that the base does not extend over the edge. Position the apparatus so that the sample channel in the wheel is visible in the microscopic field. Begin your observations to the right of the divider. Rotate the wheel counter-clockwise as you work your way through the sample.

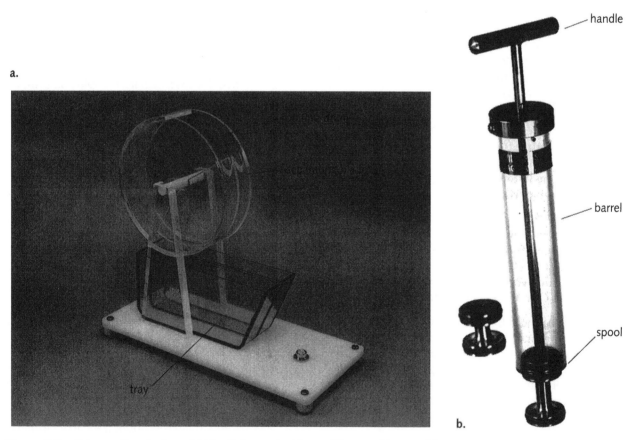

a.

b.

Figure 8.7 Subsampling devices. (a) The Folsom plankton splitter provides a quick, precise method of dividing a plankton sample into two equal parts. A measured volume of water is placed into the undivided section of the rotating acrylic splitter drum. The sample is then divided by a separating blade during 120° rotation of the drum. Rotation continues until all of the sample is poured into two deep removable trays. The contents of one tray may be returned to the splitter drum for further subdividing. (b) The Hensen-Stempel pipet.

b. Determine how many different kinds of organisms are present. Use the illustrated Keys to Plankton in this unit (page 126) and try to assign each organism to one of the taxonomic groups designated by the Keys. The Keys also identify adult and larval stages of specimens.

c. If you identify different species within any of the taxa, list them separately. For example, you might refer to different copepods as "copepod 1" and "copepod 2."

7. Count the plankton in your subsample.

a. Return to the first grid in the counting dish.

b. Count each of the different kinds of organisms and record your data in Column F of Table 8.2 on the Plankton Checklist.

c. Move to the next grid, and repeat the process until you process all grids in the dish.

d. Make the calculations for columns G (number of each planker in the collection) and H (number of each planker per cubic meter of water sampled).

$$G \text{ (Split)} = F \times 2^s$$
where s = number of splits (from Line E), or

$$G \text{ (Hensen-Stempel)} = F \times (B/St)$$
where St = volume of Stempel insert (from Line E),

$$H = G/A$$
where A is total volume of water filtered (see Step 1 of this procedure).

NOTE: These calculations will be made for the five most numerous taxa in the subsample (see Step 9 and Table 8.3). Therefore, you may not need to make these calculations for every planker observed. Consult your instructor. Refer to the sample computations that follow Step 11.

8. Return your subsample to the collection jar when you are finished.

9. Complete Table 8.3 (Summary of Data), and make your data available to the class. You may be asked to compare your observations and counts with those of other members of the class (see Step 10).

a. List the five most numerous taxa in your subsample, ranked from highest (1) to lowest (5).

b. Tally the **total plankton count** (sum of all plankters) in Column F.

c. Separate your recorded counts in Column F into **holoplanktonic** or **meroplanktonic** organisms, and total these counts. The sum of these two categories should equal the total plankton count.

d. Separate your recorded counts in Column F into: **adults,** or **eggs and larvae,** and total them. The sum of these two categories should equal the total plankton count.

e. Make the calculations for columns G **(total number of plankton in the collection)** and H **(number of each planker per square meter of water sampled).**

$$G \text{ (Split)} = F \times 2^s$$
where s = number of splits (from Line E), or

$$G \text{ (Hensen-Stempel)} = F \times (B/St)$$
where St = volume of Stempel insert (from Line E),

$$H = G/A$$
where A is total volume of water filtered (see Step 11).

See the Sample Computations that follow Step 11.

10. Complete Table 8.4. (Class Data)

11. Perform the calculations in Section B on the Plankton Analysis Report Form.

a. **Total Volume of Water Filtered:** Line A. A plankton net with a circular mouth filters a volume of water approximately equal to a cylinder whose length is the tow distance and whose radius is the radius of the net (Fig. 8.8). The information required to determine the volume can be obtained from the field data in Section A of the Report Form. Compute the **volume of a cylinder:**

$$V = \text{pi} (r^2) \times L$$

where pi = 3.1416, the radius (r) is determined from the diameter of the plankton net, and L = the distance the net is towed (in meters). The last measurement can be determined from the time of the tow and the vessel speed.

b. **Volume of Plankton/m³ of water filtered:** Line D. You measured the volume of plankton in your collection (Line C) that was present in the volume of water sampled by the collecting net (Line A). Thus,

$$D = C/A$$

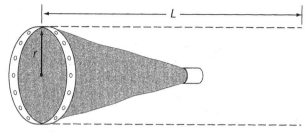

Figure 8.8 A plankton net with a circular mouth filters a cylinder of water. Determination of the volume of the cylinder reveals the quantity of water filtered.

Sample Computations for Lines A–E and Columns F–H.

A 0.5-m diameter plankton net is towed for 15 minutes at a vessel speed of 2 knots. One knot is 1 nautical mile/h; there are 1850 m in a nautical mile.

$$V = (3.1416 \times 0.25 \text{ m}^2)(2 \times 1850 \text{ m})$$

A $V = 2906 \text{ m}^3$ [volume of water filtered]

B   Total volume of the plankton collection (plankton + preservative) is 150 mL.

C   Your settled plankton volume is 50 mL.

D $= 50 \text{ mL}/2906 \text{ m}^3$
      $= 0.017 \text{ mL}/\text{m}^3$

E   You split your sample six (6) times, or your Hensen-Stempel pipet subsample is 2 mL = St.

F   You counted 384 plankters.

G   (Split) $2^s = 2^6 = 64$;
      $= F \times 2^s$
      $= 384 \times 64 = 24{,}576$ plankters in the collection, or
      (Hensen-Stempel)
      $= F \times (B/St)$
      $= 384$ plankton $\times$ (150 mL/2.2 mL)
      $= 28{,}800$ plankters in the collection.

H   The density of your plankton collection is
      (Split)
      $= 24{,}576/2906 \text{ m}^3$
      $= 8.46$ plankters per cubic meter, or
      (Hensen-Stempel)
      $= 28{,}800/2906 \text{ m}^3$
      $= 9.91$ plankters per cubic meter.

Interpretation

Prepare a report that includes the Report Forms along with a discussion of the following questions:

1. How would you describe the species diversity of this collection? You might answer this question best after calculating diversity indices for your sample. Consult Appendix II and Unit 11 for more information. On the other hand, you can make a fair appraisal of your sample by ranking the planktonic forms in order of importance based on your counts. Are there few or many different forms? Are the counts evenly distributed among the different plankters, or are one or two forms dominant?

2. Compare your results with those of other members of your class. Do major differences or variations appear in the identifications? In the numerical analyses? Try to explain any differences that you observe.

3. Identify at least two ways in which errors can occur in the original field collection of the plankton sample.

4. Identify two ways in which errors might be introduced in your procedure.

5. Were any meroplanktonic forms present? Any eggs or larval stages? To what extent? Is their presence related to the season in which the sampling occurred?

6. Are the dominant plankters present herbivores or carnivores?

7. How might the following changes affect the composition (size and species present) of the collection that you analyzed:

   a. Reduce the mesh size of the collection net to 0.150 mm.

   b. Make the collection at night.

   c. Change the sampling depth.

COMPOSITION OF PLANKTON

A. MEASUREMENT OF PLANKTON NET SAMPLES

	Net Samples		
	A	B	C
Mesh Width (mm)			

B. PLANKTON GROUPS VULNERABLE TO CAPTURE WITH EACH OBSERVED MESH SIZE
(LIST THE GROUPS AND THEIR SIZE RANGES)

	Net Samples		
Plankton Group (Size)	A	B	C

PLANKTON ANALYSIS REPORT FORM

SECTION A: FIELD DATA FOR PLANKTON COLLECTION

Date _____ Tow No. _____ Location _____

Depth of Water _____m Depth of Tow _____m

Tow Start Time _____ Tow Finish Time _____ Total Time _____

Net Diameter _____m Mesh Size _____mm Net Type _____

Vessel Speed _____kn Water Temperature _____°C

Salinity _____o/oo DO _____ppm Transparency _____m

SECTION B: QUANTITATIVE ANALYSIS OF PLANKTON COLLECTION

Line A Total volume of water filtered to obtain plankton collection (m³) = _____
From Section A, Procedure Step 11, and Fig. 8.8:
$V = \text{pi} \, (r^2) \times L$

Line B Total volume of collection = plankton + preservative (mL) = _____
(From Procedure Step 2c)

Line C Volume of settled plankton (mL) = _____
(From Procedure Step 2c)

Line D Volume of plankton/m³ water filtered = _____
From Procedure Step 10b: D = C/A

Line E Subsample: Splitter _____ Number of splits _____

Volume of Stempel pipet (mL) = _____

PLANKTON CHECKLIST

TABLE 8.2 Plankton checklist, count (Column F) of subsample (Line E), calculations of number of plankton in the collection (Column G), and per cubic meter of water filtered (Column H).

Plankton	Column		
	F Number in Subsample	G Number in Collection $F \times (B/St)$ or $F \times 2^5$	H Number per m^3 G/A

SUMMARY OF DATA

TABLE 8.3   Summary of data from Table 8.2. Count (Column F) of subsample, number of individuals in the collection (Column G), and number per cubic meter of water filtered (Column H) for
(a) the five highest-ranked plankton taxa,
(b) total plankton counted,
(c) holoplankton and meroplankton, and
(d) adults and egg/larval stages.

Category	Column F Number in Subsample	Column G Number in Collection $F \times (B/St)$ or $F \times 2^s$	Column H Number per m^3 G/A
Taxa (ranked):			
(a) Name			
1 _____	_____	_____	_____
2 _____	_____	_____	_____
3 _____	_____	_____	_____
4 _____	_____	_____	_____
5 _____	_____	_____	_____
(b) Total Plankton	_____	_____	_____
(c) All Holoplankton	_____	_____	_____
All Meroplankton	_____	_____	_____
(d) All Adults	_____	_____	_____
All Egg/Larvae	_____	_____	_____

CLASS DATA

TABLE 8.4   Class data. Means and standard deviations of class compiled data from
Table 8.3. Count (Column F) of subsample, number of individuals in the collection
(Column G), and number per cubic meter of water filtered (Column H) for
(b) total plankton counted,
(c) holoplankton and meroplankton, and
(d) adult and egg/larval stages.

	Column F Number in Subsample		Column G Number in Collection $F \times (B/St)$ or $F \times 2^s$		Column H Number per m^3 G/A	
	Mean	SD	Mean	SD	Mean	SD
Total plankton	_____	_____	_____	_____	_____	_____
$N^* =$ _____	_____	_____	_____	_____	_____	_____
(c) All Holoplankton	_____	_____	_____	_____	_____	_____
All Meroplankton	_____	_____	_____	_____	_____	_____
$N^* =$ _____	_____	_____	_____	_____	_____	_____
(d) All Adults	_____	_____	_____	_____	_____	_____
All Egg/Larvae	_____	_____	_____	_____	_____	_____
$N^* =$ _____	_____	_____	_____	_____	_____	_____

*Number of data sets used to calculate means and standard deviations.

KEYS TO PLANKTON

The principles upon which taxonomic keys are based are discussed in Unit 2. The Keys to Plankton are artificial in that they do not reflect phylogenetic relationships. Except for the initial selection from among four choices, the Keys are fundamentally dichotomous: they consist of numbered couplets, with a choice of either **a** or **b,** that lead to an identification.

To use the Keys, you must first recognize the general appearance, shape, or outline of the specimen. Your first choice leads you to other Keys and to corresponding illustrations. The Keys to Plankton include the groups of plankton that commonly occur in coastal samples: phytoplankton, holozooplankton, and larvae of benthic invertebrates (see Fig. 8.1).

As you move through the Keys, it will become evident that various taxa have a variety of shapes and sizes. Some groups (for example, diatoms) will therefore be mentioned in more than one Key. Figure 8.2 illustrates the size ranges of the forms you are likely to encounter in your sample. Phytoplankton are usually less than 1 mm in size. Most zooplankton range from 0.3 mm to 10 mm. Refer to Figure 8.2 for help with identifications.

A. Key to General Appearance

1. Spherical, globular, or round (disklike)	Key B
2. Triangular in outline, or tubular, cup, bell, or cone-shaped	Key C
3. Elliptical or ovoid in outline	Key D
4. Elongated, segmented, or chainlike; with or without an exoskeleton	Key E

B. Key to Spherical, Globular, or Disk-shaped Plankton

1 a. Spherical or globular		2
b. Disk-shaped		6
2 a. Ciliated on outer surface; cilia in conspicuous bands of plates or combs; macroscopic. *Comb jelly* (phylum Ctenophora)		
b. Not ciliated		3

3 a. With a single large tentacle undulating from ventral groove. *Noctiluca,* a dinoflagellate (division Pyrrophyta)

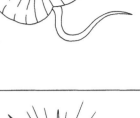

   b. No tentacles   4

4 a. With spines radiating from a spherical mass. *Radiolarian* (kingdom Protista)

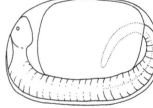

   b. Without radiating spines   5

5 a. Transparent sphere, with oil droplet, and possibly developing embryo. *Fish egg* (phylum Chordata; class Osteichthyes)

   b. Translucent, central detail lacking; in various stages of development. *Invertebrate egg*

   c. Sphere covered by calcareous plates. *Coccolith* (Coccolithophora)

6 a. Disk hard; forms a shell   7

   b. Disk soft   8

7 a. Disk siliceous, drumlike, and geometrically designed.
   Centric Diatom (division Chrysophyta; Bacillariophyceae)

   b. Disk calcareous, compartmentalized, perforated; living
   forms with cytoplasmic processes extending from
   perforations. *Foraminiferan* (kingdom Protista)

8 a. Disk is flexible, thin or flattened, transparent, with one or
   more eggs in the center. *Snail egg* (phylum Mollusca;
   class Gastropoda)

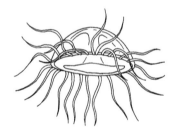

   b. Disk soft, with numerous tentacles extending from margin.
   Medusa (phylum Cnidaria)

C. Key to Triangular, Tubular, Cup- and Cone-shaped Plankton

1 a. Skeleton or test with geometric design; no appendages;
   not ciliated; microscopic; golden-brown in life   2

   b. With ciliated bands, disks, or lobes, or with conspicuous
   . appendages or tentacles   3

2 a. Tapered form, golden-brown in color. *Pennate diatom*
   (division Chrysophyta; Bacillariophyceae)

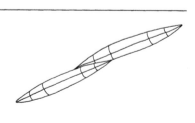

b. Minute, blunt to sharply elongated triangular form with flagella in conspicuous central girdle; some skeletons with 3 sharp spines; slow rotation when moving. *Dinoflagellate* (division Pyrrophyta; Dinophyceae)

3 a. With ciliated bands, disks, or lobes, or tentacles   4

     b. Not ciliated; body shield-shaped with 3 pairs of jointed appendages. *Nauplius larva.* See also Key C, 7a.   7

4 a. With ciliated bands, disks, or lobes   5

     b. Body with tentacles   6

5 a. Cone-shaped, with paired, ciliated, lobes, or disks. a. *Pilidium larva of worm* (phylum Nemertea), b. *Cyphonautes larva* (phylum Bryozoa)

     b. Cone-shaped, with conspicuous band of elongated cilia near anterior end. *Trochophore larva of worm* (phylum Annelida; class Polychaeta)

     c. Tubular; cilia encircle one end of cell in transparent tubular cone (lorica). *Tintinnid* (kingdom Protista)

6 a. Cone-shaped; single tentacular appendage may be evident. *Siphonophore* (phylum Cnidaria)

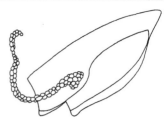

   b. Cone-shaped; 4 or more tentacles extending from conspicuous head with well-developed eyes. *Squid* (phylum Mollusca; class Cephalopoda)

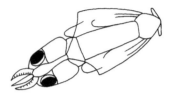

   c. Bell-shaped or inverted cup-shaped with marginal tentacles. *Jellyfish;* may be *medusa* or *ephyra* stage in life cycle (phylum Cnidaria). See also Key B, 8b.

7 a. Shield bears "horns" projecting from the anterior corners. *Nauplius of barnacle* (phylum Arthropoda; class Crustacea; order Cirripedia)

   b. No such projections of body shield. *Nauplius of copepod* (class Crustacea; order Copepoda). See Key C, 3b.

D. Key to Elliptical Ovoid Plankton

1 a. Shell or shell-like body covering	2
b. Without shell or shell-like carapace	9
2 a. Bivalve (2 shells) appearance; may enclose segmented body	3
b. Univalve (one shell), or body segmented	4

3 a. Clamlike; no jointed appendages; may possess ciliated bands and elongated foot. *Larval clam* or *Veliger larva of bivalve mollusk* (phylum Mollusca; class Bivalvia)

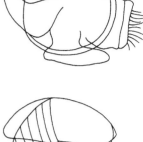

b. Tear-drop shape; may show jointed appendages, antennae, and segmented abdomen. *Cypris larva of barnacle* (class Crustacea; order Cirripedia)

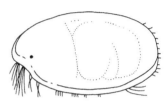

c. Bean-shaped; appendages may extend out from beneath the shells; eggs often visible. *Ostracod* (class Crustacea; order Ostracoda)

---

4 a. Body segmented; jointed appendages evident (class Crustacea; several orders follow) 5

b. Snail-like shell, coiled, or with several compartments 7

---

5 a. With large, single, anterior compound eye. *Cladoceran* (class Crustacea; order Cladocera)

b. Lacking such an eye 6

---

6 a. Elliptical, or tear-drop shape; 3 pairs of appendages; segmentation may not be conspicuous. *Nauplius larva of barnacle, copepod,* or *copepodid of copepod.* See also Key C, 3b and 7a.

b. Conspicuous segmentation; body flattened dorso-ventrally; several pairs of similar jointed appendages. *Isopod, sow bug* (class Crustacea; order Isopoda)

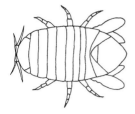

c. Conspicuous segmentation; body laterally compressed. *Scud, amphipod* (class Crustacea; order Amphipoda)

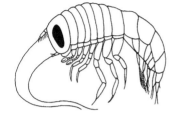

7 a. Internal organs evident; shell entire   8

b. Uncoiled shell with compartments, perforated, usually empty. *Foraminiferan* (kingdom Protista)

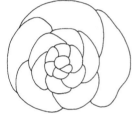

8 a. Foot modified to resemble paired wings. *Pteropod* (phylum Mollusca; class Gastropoda)

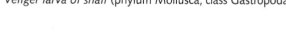

b. Shell variously shaped, expanded ciliated folds on foot. *Veliger larva of snail* (phylum Mollusca; class Gastropoda)

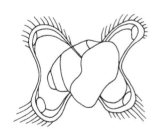

9 a. Barrel-shaped, with conspicuous circular bands   10

b. Departure from barrel-shape, with external ciliated bands and evidence of internal gut   11

10 a. Transparent, tubular with circular bands of muscle. *Salp*
   (phylum Urochordata)

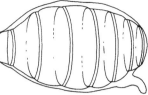

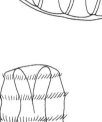

   b. Transparent, tubular with external circular bands of cilia.
   Pentacula larva of sea cucumber (phylum Echinodermata;
   class Holothuroidea)

   c. Slight C-shape in lateral view, with ciliated lateral bands.
   Bipinnaria larva of seastar, Auricularia larva of sea cucumber
   (phylum Echinodermata; class Asteroidea or Holothuroidea)

11 a. Tapered, blunt end with ciliated band; tapered end may be
   segmented; setae present. *Trochophore or early larva of worm*
   (phylum Annelida; class Polychaeta)

   b. Elongated ciliated "arms" or extensions radiate from central
   body mass. *Ophiopluteus larva of serpent star* (phylum
   Echinodermata; class Ophiuroidea) or *Echinopluteus of sea urchin*
   (phylum Echinodermata; class Echinoidea)

E. Key to Elongated and Segmented Plankton

1 a. Chainlike repetition of geometrically shaped units, golden-
   brown color. *Centric diatoms* (division Chrysophyta)

   b. Body segmented   2

2 a. Bristles (setae) present. *Polychaete worm, nectochaete
   larvae* (phylum Annelida; class Polychaeta). See Key D, 11a.

   b. No bristles or setae, or, if present, confined to anterior
   region of jaws   3

3 a. Exoskeleton and jointed appendages (class Crustacea;
   several orders)   4

   b. Without jointed appendages; lacks exoskeleton   11

4 a. Shieldlike body with 3 pairs of appendages. *Nauplius or
   copepodid larva of copepod.* See Key C, 3b and
   Key D, 6a.

   b. With more than 3 pairs of appendages   5

5 a. Shrimplike in appearance   6

   b. Not shrimplike   8

6 a. Carapace covers entire thorax   7

   b. Carapace does not cover entire thorax; statocyst on tail-
   fan. *Mysid, opossum shrimp* (order Mysidacea)

7 a. Stout abdomen with appendages; tail-fan with telson and
   uropods. *Shrimp, prawn* (order Caridea or Penaeidea)

   b. Slender abdomen without appendages; tail-fan of telson
   only with terminal spines. *Zoea larva of shrimp or prawn*
   (order Caridea or Peneidae)

8 a. Crablike in appearance. *Megalopa larva of brachyuran crab*
   (order Brachyura)

   b. Not crablike, but with enlarged cephalothorax   9

9 a. With single, anterior eye; paired antennae extending
      laterally. *Copepod (order Copepoda)*

   b. Not as above   10

---

10 a. Large, bulbous thorax with elongated abdomen, void of
       appendages. *Cumacean (order Cumacea)*

    b. Carapace with exaggerated anterior, posterior, and (in some
       species) dorsal spines. *Zoea larva of brachyuran crabs*
       (Brachyura)

---

11 a. With supporting rod (notochord) or appearance of segments
       along dorsal body wall   12

    b. Transparent, arrowlike, with fins; anterior jaws with chaetae.
       Arrow worm (phylum Chaetognatha)

---

12 a. With typical vertebrate structure, paired eyes, fins, yolk sac.
       Fish larva (class Osteichthyes)

    b. Enlarged head, tail-like structure with notochord,
       segmentation, tadpole-like. a. *Ammocoetes larva of tunicate*
       (Urochordata), b. *larvacean* (Urochordata)

Fish Morphology and Diversity

OBJECTIVES

After completing this unit, you will be able to

- Recognize the basic external morphology of fishes representing the classes Chondrichthyes and Osteichthyes;

- Recognize key characters for the identification of marine fishes;

- Identify differences among major orders of fishes; and

- Summarize natural history information for designated orders, families, and species.

INTRODUCTION

Nekton are large, actively swimming pelagic animals. Five thousand such species exist, although they include few taxonomic groups. Squids and a few species of shrimps are the only invertebrates that are truly nektonic. Most nekton are vertebrates (phylum Chordata; subphylum Vertebrata), and most of these are fishes.* Other nektonic marine vertebrates include sea turtles and sea snakes (class Reptilia), and seals and whales (class Mammalia).

Fish are ectothermic vertebrates that typically have dermal scales, external gill openings, external nostrils (nares), mouth, muscular trunk, and median fins. Living fish are classified into three classes based on the structure of the jaw and the degree of calcification of the skeleton. Jawless fishes **(class Agnatha)** include lampreys and hagfishes. Cartilaginous fishes (sharks, skates, and rays) belong to the **class Chondrichthyes,** an old and successful group comprising relatively few species. The **class Osteichthyes** (bony fishes) represents the largest group of vertebrates both in number of species (at least 20,000) and in number of individuals. By adaptive radiation these fishes have developed an amazing variety of forms and structures. They flourish in freshwater and seawater, and in both deep and shallow water.

This unit introduces you to the basic external morphology of two classes of extant (living) fishes. The information in this section is necessary to understand and use the Key to Orders of Fishes in this unit.

A. External Morphology

In this portion of the unit, you will study basic external morphology of examples of sharks, skates, rays, and bony fishes. You may have the opportunity to compare the morphology of several different specimens for each feature described. Your familiarity with the characters presented will facilitate your use of the Key in part B.

*The word *"fishes"* refers to more than one species; *"fish"* can be plural, but it means more than one individual, regardless of species.

GENERAL PROCEDURE

1. Become familiar with this basic set of terms that define anatomical position and direction (see Fig. 9.1). Notice that they express the relative position of one structure to another.

   Cranial (G. *Kranion,* head)—pertaining to the cranium (skull), the front or head end; *see* **anterior.** "Your belly button is cranial to your buttocks."

Caudal (*L. Cauda,* tail)—pertaining to the tail or region of the tail; *see* **posterior.** "Your buttocks are caudal to your belly button."

Anterior (*L. Foremost*)—pertaining to the front or head end; *see* **cranial.** "Your belly button is anterior to your buttocks."

Posterior (*L. Latter*)—the tail, opposite of anterior; *see* **caudal.** "Your buttocks are posterior to your belly button."

Dorsal (*L. Dorsum*)—pertaining to the back or upper surface. "Your backbone is dorsal to your belly button."

Ventral (*L. Venter,* belly)—lower side or belly; away from the back; opposite of dorsal. "Your belly button is ventral to your backbone."

Medial (*L. Medius,* middle)—the midline or near the middle of the body. "Your belly button is medial in position." "Your sternum is medial to your arm."

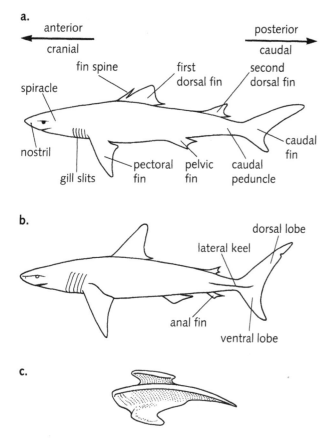

Figure 9.1 External morphology of sharks. (a) Dogfish shark with dorsal fin spines. (b) Mako shark with anal fin and lateral keel on caudal peduncle. (c) Placoid scale.

Lateral (*L. Latus,* side)—toward the side of the body; at each side of the median line. "Your arm is lateral to your sternum."

2. Obtain a large, metal dissecting tray, cover it with paper towels, and then obtain a fish specimen.

CAUTION

You may be working with specimens preserved in formalin, which is an irritant and is potentially carcinogenic. Wear protective gloves and work in a well-ventilated space. No special precautions are necessary when working with specimens stored in alcohol.

NOTE: You may study relatively large preserved specimens that are customarily stored in large containers. It is important to keep these specimens *wet!* Moisture is particularly critical to maintaining the integrity of the fins, which dry quickly. Return your specimen to the storage container when you finish your observations.

3. Locate the features described in the following exercises, refer to the accompanying illustrations, and make notes and sketches as you work.

   a. The questions asked in the procedures should aid you in your observations. Attempt to answer them in your notebook.

   b. Make your drawings in your notebook. Be certain to label the drawings and make appropriate notations.

4. Discard any biological waste, including the carcass, in the appropriate containers. Waste from fresh specimens must never be mixed with preserved waste. Waste toweling must not be mixed with biological waste.

E X E R C I S E   1

Cartilaginous Fishes (Class Chondrichthyes)

Members of class Chondrichthyes (*chondr* = cartilage; *ichthys* = fish)(Fig. 9.1) have a well-developed **lower jaw, bony teeth** on both jaws, **paired fins, gills** with individual external openings, a pair of **spiracles** (openings on the top of the head that lead to the gill chambers), **claspers** (elongated modifications of the ventral fins in mature males), and **placoid scales.**

The class is divided into orders that are easily distinguished by form. The descriptions that follow begin with the highly publicized and well-recognized shark (Fig. 9.1a,b). Skates and rays (Fig. 9.2) are presented as modifications of the shark form.

Sharks

The body form of most sharks is **fusiform**—that is, somewhat torpedo-shaped and ovoid in cross section (Figs. 9.1a,b). Two sets of paired fins exist: the larger and anterior-most **pectorals** and the smaller, posterior **pelvics.** The latter pair of fins are modified in mature males, where they possess an elongated process **(clasper)** (see Fig. 9.2a) that functions as a copulatory organ to ensure internal fertilization of the female. Most sharks have two **dorsal fins,** which may be accompanied by a **spine** (Fig. 9.1a). A single **anal fin** (Fig. 9.1b) may be positioned between the paired pelvics and the **caudal fin.** The caudal fin in sharks is **heterocercal** in shape; the dorsal lobe is larger than the ventral lobe. This shape creates a net downward thrust that is offset by the enlarged pectoral fins, which are used primarily as horizontal stabilizers rather than for propulsion.

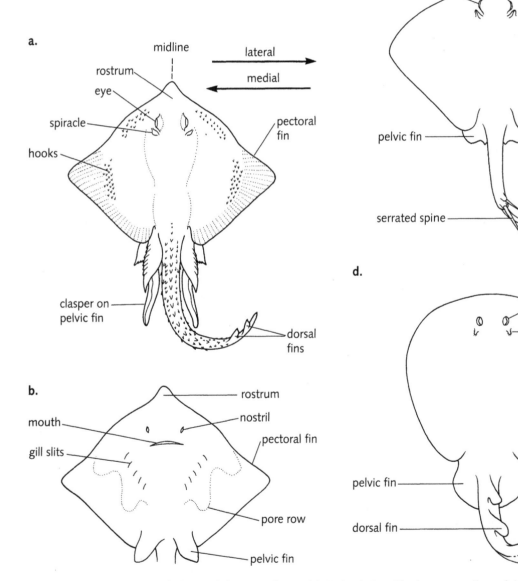

Figure 9.2 External morphology of skates and rays. (a) Male skate with claspers on the pelvic fins (dorsal view). (b) Ventral view. (c) Stingray. (d) Torpedo ray.

Five to seven **gill slits** can be found on each side of the head directly anterior to the pectoral fins. Each gill slit is associated with a single gill. A pair of **spiracles** located dorsally on the head open into the gill chambers.

1. Examine the spiracles. Is there any evidence of valves?

2. Find the **nares** (= nostrils) and the **eyes.** How would you describe the location of these sensory organs? Anterior? Lateral? Dorsal? What is the shape of the pupil of the eye?

3. Examine the surface of the head closely, perhaps under a dissecting microscope or with a hand lens. Look for linear series of pores in the skin that form a pattern over the head and extend caudally along the lateral side of the trunk to the tail. These pores are part of the **lateral line** system, a tubular system that contains sensory organs receptive to pressure and temperature changes. Sketch the pattern of the sensory pores.

4. Gently rub your fingers along the surface of the shark in a caudal direction (toward the tail). Now rub in the opposite direction (cranial—toward the head). Notice the difference in texture.

5. Examine the skin, or a prepared slide of shark skin, under the dissecting microscope. The skin's roughness is due to the caudally directed spines of **placoid scales** (Fig. 9.1c), characteristic of the Chondrichthyes. They are extensively modified in some species, but are missing in others. Draw a scale from your specimen.

6. Placoid scales are developmentally and histologically similar to the **teeth.** Most sharks possess a triangular-shaped dentition that is continually replaced throughout the life of the individual. Tooth shape varies, ranging from the very large and serrated variety to small, platelike forms. Are the teeth in your specimen uniform in size and shape? Draw a few teeth from your specimen.

Skates and Rays

These two groups make up the cartilaginous "flat fishes." The dorso-ventrally flattened body (**depressiform**) (Figs. 9.2 and 9.3d) of the flat fish is intimately associated with a life on the sea floor (manta rays are an exception). The shape of skates and rays largely derives from the greatly enlarged **pectoral fins** that extend from the head to the origin of the **pelvic fins.** Other modifications of the shark form accompany the depressed condition.

Skates and rays are not alike. Several taxonomic orders and families of rays exist, while all skates are members of the family Rajidae. Skates differ from rays primarily by having a relatively short tail with two dorsal fins. In skates, the pectoral "wings" also extend forward around a thin plate of cartilage that forms a prominent **rostrum** (Fig. 9.2a,b).

1. Notice the location of the **mouth** and **gill slits;** they are on the ventral side of the animal (Fig. 9.2b). The **spiracles,** which are more dorsal in position than they are on the sharks, are the principal mechanism for producing the flow of clean water over the gills. Spent (deoxygenated) water is expelled through the ventral gill slits.

2. The **pectoral fins** produce the main propulsive force for locomotion, and the trunk musculature (caudal to the pectoral fins) is reduced, as is the **caudal fin.** Note the greatly reduced size of the **dorsal fins.** Examine the **pelvic fins** for evidence of **claspers** (Fig. 9.2a). Is your specimen male or female?

3. Examine the mouth and teeth. Consider the shape of the teeth and the body form. Based on these characters, what can you deduce about the diet of the animal? Sketch the teeth.

4. Examine the skin of a skate or ray. Placoid scales may be either absent or modified as small dermal hooks, thornlike tubercles, or spines. The presence and location of these specialized placoid scales are used in identification. Compare the shape of spines with that of the shark's placoid scale. Can you detect any regular pattern on the skin of your specimen? Sketch the spination on your specimen.

   Most rays—except the family Torpedinidae (see Fig. 9.2d)—are called stingrays because of the long **serrated spine** (or spines) projecting dorsally from the base of the tail, near the dorsal fin (Fig. 9.2c). Glandular tissue associated with the spine secretes a venom. The spine, which is a derivative of a placoid scale, contrasts with the smooth body surface.

5. Examine a spine under magnification. Is the spine potentially more useful as an offensive or defensive weapon? Would the spine leave a "clean" wound? Sketch a stingray spine.

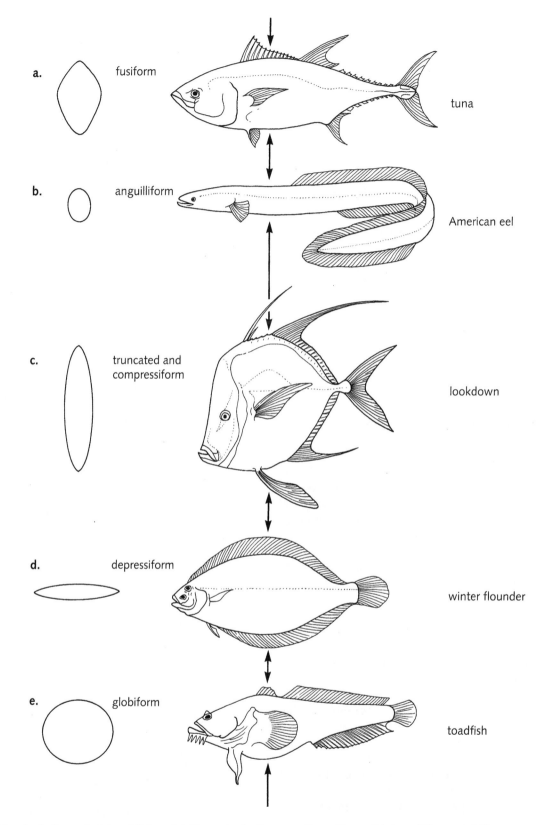

Figure 9.3 Body forms in teleost fishes. (a) Elongate, fusiform, or thunniform. (b) Anguilliform, eel-like, or attenuated. (c) Ovate (truncated), and compressiform, compressed (thin, narrow, deep). (d) Depressiform, depressed, or flattened. (e) Globiform.

6. Examine the ventral surface of the animal. What is the adaptive advantage of its white color? Look for evidence of the **sensory pore** system, which is similar in design and function to the lateral line (see description in Bony Fishes). Explain why you would or would not expect a skate or ray to have a sensory pore system on the ventral side.

Interpretation

1. Describe the anatomical positions of the nares and eyes in the shark. Relate their locations to shark behavior.

2. Relate dentition with diet of the specimens you examined.

3. What is the adaptive advantage of a white ventral surface on skates and rays? On sharks?

4. Relate the presence or absence, and the location, of the lateral line system to habitat. Contrast, for example, skates and sharks.

E X E R C I S E   2

Bony Fishes (Class Osteichthyes)

The class Osteichthyes (*oste* = bone; *ichthys* = fish) is a highly diverse taxon, making it impractical to describe all the morphological variations in this exercise. You can appreciate such variations by comparing and contrasting your observations of different specimens with the morphology described in this manual.

1. Describe the shape of your specimen. The body is typically torpedo-shaped (**fusiform,** Fig. 9.3a) and slightly to strongly ovoid in cross section. Many fishes depart from this generalized shape, and many terms have been derived to describe the variations: **anguilliform** (Fig. 9.3b; eel-like, attenuated), **ovate** or **truncated** (Fig. 9.3c), **compressiform** (Fig. 9.3c; flattened from side to side, thin, perchlike), **depressiform** (Fig. 9.3d; flattened from top to bottom), and **globiform** (Fig. 9.3e; subcircular). More than one term may be used to describe a fish (for example, the lookdown is truncate and compressiform).

   Draw lateral and cross-sectional outlines of your specimen. Complete the drawing of the lateral view as you identify the external features described below.

2. Identify the **pectoral, pelvic, anal, dorsal,** and **caudal fins** (Fig. 9.4a). How many of each fin are there? Are any missing? Which occur in pairs? Are single (unpaired) fins continuous, or are they divided either partially or completely into two separate structures?

Is an **adipose fin** (Fig. 9.4b) present? This small, median, fleshy (no supporting elements) dorsal fin lies near the caudal fin. Adipose fins occur in the more primitive fishes, such as salmon.

Where are the pelvic fins located in your specimen? They are more abdominal in position in more primitive (evolutionarily) fishes, such as herring and salmonids (Fig. 9.4b). In more advanced fishes, such as bass, the pelvics are thoracic in position, near the pectorals (Fig. 9.4a).

3. Examine the construction of a dorsal and pectoral fin with and without a dissecting microscope. Note the **fin rays** (Fig. 9.5), the supporting elements of the fin. They can be composed of **spines, soft rays, spinous soft rays** (also called hard rays), or a combination thereof. **Spines** have a single shaft, and are hard and sharp; they may stand free or connect to other rays by tissue. **Soft rays** are bilateral, having two halves to the shaft. They are often segmented, and branched or tufted. **Spinous soft rays** are bilateral, but hard and unbranched.

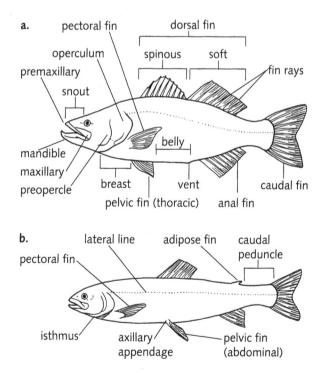

Figure 9.4 External features of bony fish. (a) Spiny-rayed, basslike fish; pelvic fins thoracic in position. (b) Soft-rayed, salmonidlike fish; adipose fin present; pelvic fins abdominal in position.

What advantage might soft rays provide for a fish? Spinous rays?

4. Count the number of spines and rays in the fins of your fish. Such counts, called **meristic characters,** are used as diagnostic characters in fish classification. Spines are recorded by Roman numerals, soft rays by Arabic numbers. For example, "Anal III,5" indicates a count of three spines and five soft rays in the anal fin. Record the counts in your notebook and compare them with other counts made in your class. Do counts vary among different specimens of the same species?

5. Describe the shape of the caudal fin (Fig. 9.6). Are **finlets** present? This series of small median fins is located both dorsally and ventrally in front of the caudal fin (Fig. 9.6f). They occur on fast-swimming fishes, presumably reducing drag created by water currents flowing over the body during swimming. If your specimen has finlets, then it likely has a forked tail supported by a relatively narrow **caudal peduncle.** It may also have **lateral keels**—fleshy, horizontal flanges that serve as stabilizers during swimming—on the caudal peduncle.

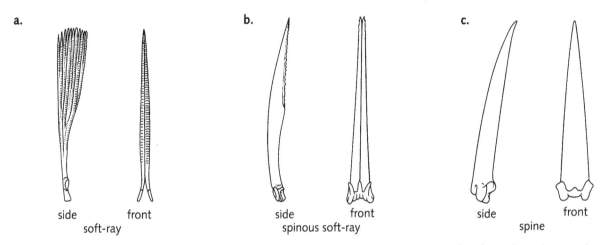

a.
side   front
soft-ray

b.
side   front
spinous soft-ray

c.
side   front
spine

Figure 9.5 (a) Soft ray and (b) spinous soft ray showing bilateral structure. Soft ray also shows branching and segmentation. (c) The spine is a single element.

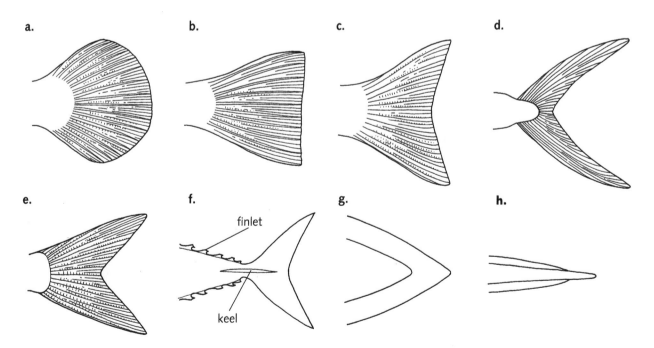

a.

b.

c.

d.

e.

f.
finlet
keel

g.

h.

Figure 9.6 Various shapes of caudal fins of bony fishes. (a) Round. (b) Square, truncate, or straight. (c) Emarginate. (d) Lunate. (e) Forked. (f) Forked, with finlets and keels on caudal peduncle. (g) Pointed or tapered, with caudal fin not differentiated from anal or dorsal fins. (h) Naked tail, no fin rays on tip.

6. Measure your specimen. The following dimensions, illustrated in Fig. 9.7, are used routinely in fishery biology and taxonomy. All are straight-line distances.

Total length (TL)—from the tip of the snout to the tip of the caudal fin (subject to error if the fin is damaged).

Fork length (FL)—from the tip of the snout to the base of the fork of the caudal fin (the most reliable body length measurement).

Standard length (SL)—from the tip of the snout to the base of the caudal fin (subject to error if the point of the base cannot be identified accurately).

Head length (HL)—from the tip of the snout to posterior-most point of opercular membrane.

Body depth (BD)—the greatest vertical distance, exclusive of fin rays or any fleshy or scaly structure, measured at the deepest part of the body (best measured with Vernier calipers).

Body width (BW) (not shown in Figure 9.7)—the greatest horizontal distance across the thickest part of the body (best measured with Vernier calipers).

Fin length (or **width**)—the straight-line distance across the base of the fin.

a. Record these measurements in your notebook, and calculate the following ratios, also used in fish identification:

Body depth to length, expressed as either BD/TL or TL/BD. Other lengths can be substituted for TL.

Body width to length, expressed as either BW/TL or TL/BW.

b. Relate these ratios to the body shape that you described for your specimen.

7. Does your specimen have **scales?** Are you certain? Do not assume that all fish have scales. On the other hand, do not assume that your specimen lacks scales because they are not obvious to you. They may have sloughed off during capture and handling, or they may be minute and embedded in the skin.

a. Examine the surface of the animal with a dissection microscope. If scales are present, note their arrangement (Fig. 9.7). Follow a row of scales from the mid-dorsal line to the mid-ventral line. Does the row run in a vertical or an oblique direction?

b. Are the scales large or small? Scale counts **(meristic)** are used as diagnostic characters.
   1. Count the number of scales in a diagonal row on one side of the body from the mid-dorsal line to the mid-ventral line. How many lie above the lateral line? How many below?
   2. Count the number of scales along the lateral line. Begin with the scale touching the pectoral fin base. End the count at the

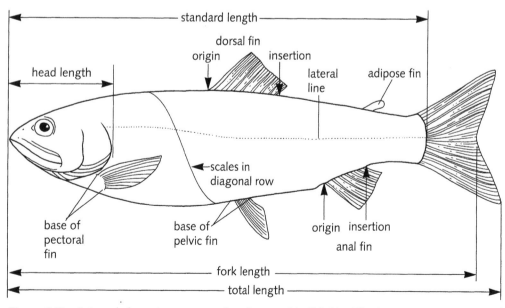

Figure 9.7 External characters commonly referenced in fish identification. Measurements and scale counts (diagonal row and lateral line).

base of the caudal fin. Does the lateral line extend onto the caudal fin surface?

3. Compare your count with those taken on other species in the class.

c. Remove a scale from the region between the anterior part of the dorsal fin and the lateral line. You may have to use forceps if the scales are deeply embedded in the skin.

1. If the scales are large, clean them by dipping them in fresh water and rubbing them between the thumb and fingers.

2. Clean smaller scales with fine forceps, brushes, and alcohol.

d. Mount the cleaned scales between two glass slides and tape the slides together. Examine them under a dissecting microscope.

1. With the aid of Fig. 9.8, note whether the scale is **ctenoid** (has small marginal spines) on it (which margin?), or if it is smooth (**cycloid**) or **ganoid** (rhombic shape). Scale type is correlated with the presence or absence of spiny rays in the fins. What association have you observed for your specimen? Draw a scale and label the features.

2. Identify the **focus** (the center of origin of the scale), **circuli** (concentric rings on the outer scale surface), and **annuli** (zones of crowded circuli that can often be followed around the scale) (Fig. 9.8d). Annuli are clusters of circuli that appear to be broken or to cross over one another.

3. Annuli identify the annual increase in size of the scale (and presumably indicate the age of the fish). Try to determine the age of your specimen.

8. Examine the **mouth** (Fig. 9.9) of your specimen. The mouth is **terminal** in position if it is directly at the front of the body. Alternatively, it may be **superior** (opening dorsally because of projecting lower jaw) or **inferior** (slightly to prominently overhung by the snout).

The mouth is bordered by lips that may be cartilaginous or membranous, or variously fleshy and often **papillose** (covered with nipplelike projections). Normally the lips are scaleless and have minute sensory organs. Grazers and suctorial feeders have specially developed lips and adaptations of other mouth parts.

The mouth ranges from small to huge (in relative size), and varies in shape. Feeding may occur by suction through an elongated snout, or by selective grazing with sharp teeth in an elongated beak. Some predators can form temporary tubes in which to engulf their prey by forward extension of the jaws. Other adaptations of skull bones and articulations increase the gape of the mouth. Move the jaws of your specimen. Based on your observations, deduce the feeding habit of the specimen.

Just behind the lips, across the front of both the upper and lower jaw, are membranous **oral valves.** These transverse flaps of tissue prevent the outflow of water during respiration (ventilation). Make a drawing that shows the position and gape of the mouth.

9. Open the mouth and examine the **teeth.** They are classified according to position (Fig. 9.10): on the upper jaw (**premaxilla** and **maxilla**), roof of the mouth (**vomer**), lower jaw (**mandible**), the **tongue,** and in the **pharynx.**

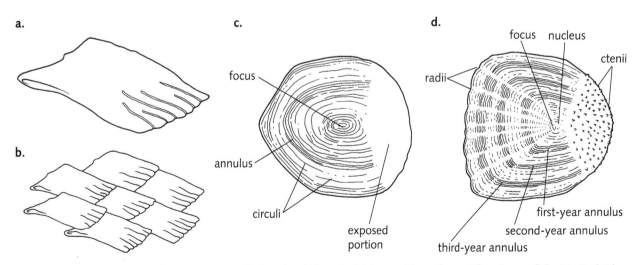

Figure 9.8 Scales of bony fishes. (a) Ganoid (rhomboid shape). (b) Disposition of ganoid scales on fish. (c) Cycloid. (d) Ctenoid showing groups of concentric rings (annuli) and other characteristic features.

Pharyngeal teeth (Fig. 9.10c), when present, occur in upper and lower sets. Their form varies according to feeding habit.

Based on form (Fig 9.11), teeth are classified as **canine** (dogtoothlike, elongated and subconical, straight or curved), **incisor** (sharply cutting), **molariform** (flattened, grinding or crushing), **cardiform** (numerous, short, fine, pointed), and **villiform** (elongated cardiform). A strong correlation exists between the form of dentition, feeding mechanism, and food eaten. Make a drawing that shows the position and kind of teeth in your fish.

10. Lift the bony plate **(operculum)** that extends caudally from the eyes (Fig. 9.4). An **opercular membrane** extends from the posterior and ventral margins. The ventral portion of the membrane is supported by bony elements **(branchiostegal rays)** (Fig. 9.12a). The membrane fits snugly against the body to close the branchial cavity during part of the respiratory cycle. Are the head and operculum in your fish ornamented or armed with spines?

11. One or two **nares** (Fig. 9.12) on each side of the snout typically lead to a blind sac lined with olfactory epithelium. Examine them under magnification, and describe their number and location. Known relationships exist between the degree of development of the nostrils and water clarity and the development of visual sense. Can you deduce what those relationships might be?

12. Observe the **eyes.** What shape does the iris have? Is there any evidence of lids or membranes covering the eyes?

   a. Measure the eye diameter. Express eye size as the ratio between eye diameter and head length (see Fig. 9.7). Would you describe the eyes as small or large? Compare your values

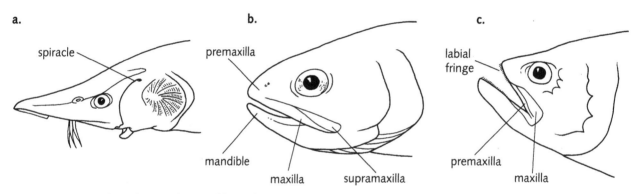

Figure 9.9 Mouth and snout forms of bony fishes. (a) Snout overhanging inferior or ventral mouth of sturgeon. (b) Terminal mouth of trout. (c) Superior mouth of sandfish.

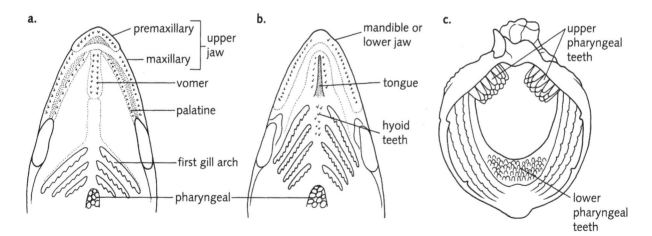

Figure 9.10 Dentition in bony fishes: location. (a) On bones of upper jaw, roof of mouth, and in pharynx. (b) On tongue, on bones of floor of mouth, and in pharynx. (c) Anterior view of gill arches and pharyngeal teeth of surfperch.

with those calculated by other members of your class.

b. Describe the position of the eyes on the head. Are they on the dorsal surface and facing upward (superior), or on the side of the head (lateral), or in some other position? Estimate the angle of their line of sight forward. Speculate as to whether your specimen has stereoscopic vision. Defend your position.

13. Numerous pores occur on the lateral surface of the body; they reveal the presence of a line that is often curved (Fig. 9.4). Remove a scale from this **lateral line** and examine it microscopically. The "line" that you can see with the naked eye is a row of small pores or tubules in the scales that connect with a long tubular canal in the skin (dermis). The canal bears ciliated sensory organs that are sensitive to pressure and temperature changes; they are also responsive to water currents. Draw a lateral line scale near your previous drawing of a normal scale. Compare the two.

Interpretation

If two or more species were observed, compare and contrast them in your answers.

1. Calculate the ratios of body depth, and body width to length, and relate these to the body form you designated for your specimen.

2. How would these ratios differ for fishes with other body shapes? In other words, select two body shapes from Figure 9.3 that differ from your specimen, and estimate the ratios.

3. Relate body form and fin morphology of your specimen to potential swimming speed and maneuverability.

4. Describe how the fins are used during swimming. Base your answer on observations of living specimens. Consider the following in your answer: What muscles and fins are used for propulsion? For maneuverability? For maintaining position? Indicate whether you observe different patterns for different species.

5. Does fin use vary with body form?

6. Body form in fish did not evolve randomly: It is highly correlated to the habitat, niche, and swimming ability of the species. Relate the shape of your specimen to its swimming ability and lifestyle.

7. What is the association between scale type and the presence or absence of spines and spinous rays in the fins?

8. What adaptive advantages do scales provide?

9. Could you use lateral line scales for age and growth studies?

10. What advantage might soft fin rays provide? Spinous rays?

11. Describe the probable food and feeding habit of your specimen. Defend your answer by referring to your observations of body form, mouth type, and dentition.

12. Does your fish have binocular (stereoscopic) vision? Defend your answer.

13. Relate development of the nares and eye to water clarity.

14. Is your specimen relatively advanced evolutionarily, or primitive? Base your answer on your observations of the relative position of the paired fins.

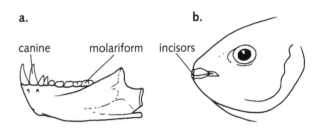

Figure 9.11 Dentition in bony fishes: form. (a) Mandible with canine and molariform teeth. (b) Fused beaklike incisors of parrotfish.

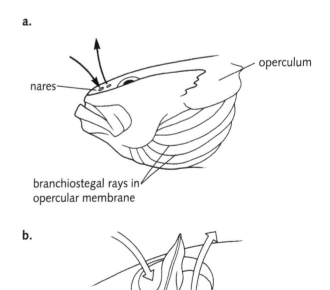

Figure 9.12 Head morphology. (a) Lateral view of head showing position of nares and the gill membrane supported by branchiostegal rays. (b) Typical naris divided by flap of skin. Arrows indicate direction of water flow.

B. Fish Diversity

Worldwide distribution and relative abundance of different groups of fishes provide some measure of their contribution to commercial fishery landings. Fisherman catch clupeoid fishes (order Clupeiformes) in the greatest numbers. Anchovies, pilchards, sardines, and menhaden are converted into fish meal for poultry and livestock feed; sardines and herring are canned or consumed directly. Most of the second-ranked gadoid fishes (order Gadiformes), such as cod and haddock, are consumed directly as fresh fish or as frozen fillets or fish sticks. Some—for example, hakes—may be converted into fish protein concentrate. The scombroids (order Perciformes, family Scombridae) are the third most important group. Mackerels account for more than half of the scombroid catch; the remainder consists of tunas. Ranking behind the top three groups are other families (jacks, basses, flatfishes, salmonids, drums) of the order Perciformes, cartilaginous sharks, skates, and rays.

Accurate identification of fishes and a knowledge of their distribution, habits, and habitats are important to fishermen and to amateur and professional ichthyologists. Identification of unknown specimens requires the orderly and efficient use of information about morphological characters and how they vary among species and other fish taxa. Many of these characters were described earlier in this unit, in the discussion of external morphology. Unique characters used to identify members of a fish group are called **key characters** for that group.

Classification, nomenclature, identification, and the use of taxonomic keys are described in the introduction to Unit 2 (Marine Invertebrate Classification and Identification). Review those concepts before you continue with this unit.

E X E R C I S E 1

Using Taxonomic Keys to Identify Orders of Fishes

Taxonomic keys are designed primarily to identify specimens by following the easiest and fastest path possible. Most are **artificial** in that the distinguishing characters used, and the order in which the forms are treated, do not indicate phylogeny. Some keys may reveal some phylogenetic classification by virtue of placing closely related species near each other. Many excellent characters used to measure phylogenetic relationships are not easily observed

(for example, internal skeletal patterns) and thus are not good key characters.

Most keys are designed to identify species within a limited geographic region. Such keys may cause misidentifications if applied to species outside the intended region.

The number of orders and families, and their names, vary among references. The variation reflects either the creation of new families (known as *splitting* by taxonomists) or the collapse of families into a single family (called *lumping*). Two standard references worth consulting are Nelson (1994) and Robins et al. (1980).

Although a taxonomic key is a shortcut that reduces the need to read copious amounts of descriptive material or to examine every reference specimen in the archives, the process of identification requires *both keying and verification*. You should first identify the fish as belonging to an order or family in a **preliminary key,** such as those used in this exercise. When you know the family, you can *tentatively* identify the species by using another, more detailed key to genera and species.

You must ultimately check the tentative identification by (1) **verifying** that the collection site is within the known range of the species distribution, and (2) **comparing** your specimen with published illustrations, descriptions, and reference specimens.

PROCEDURE

1. Fresh, preserved (in formalin, ethanol, or isopropanol), or living specimens will be available for identification. You will be instructed which species to select, and the appropriate reference keys to use for each.

CAUTION

Formalin is an irritant and is potentially carcinogenic. Wear protective gloves if your specimens are preserved in this reagent.

   a. Use metal pans or dissecting trays for freshly collected, or preserved specimens. Keep the specimens *wet!*

   b. Living fishes can be identified *in situ* (in the aquarium) or in smaller tanks that you may take to your study space. In the latter case, avoid keeping the fish in the smaller container for a prolonged period.

Do not handle living specimens with hands or gloves that have been contaminated with any preservative.

2. Work in pairs. Key each specimen to **order** by using the **Key** in this unit.

   a. In your first try, work with a known specimen and backtrack through the key.

   b. Now try an unknown. Do not skip couplets. Read both statements of the couplet before you select one. Be sure that you understand each key character and its contrasting states.
      1. If you are uncertain of the meaning of a term or character, refer to descriptions and figures in the external morphology section and to glossaries in guides or other keys.
      2. In your notebook, list the sequence of couplets that you follow. You can then backtrack if you make a mistake.

   c. If possible, work with more than one specimen before making a decision. Specimens can be damaged, and species vary within character states.

3. Obtain copies of other keys or references that permit identification of species, and identify the fish to the lowest possible taxonomic unit listed. Keep records of the keys used in your identifications.

4. Verify your final determination:

   a. Match your specimen to a reference description of morphology and size.

   b. Assure yourself that your specimen looks like the illustration.

   c. Be certain that the specimen's capture locality lies within the known range of the species.

   d. Have your instructor verify the identification.

NOTE: If you made an error, you may not need to return to the beginning of the key. Carefully work backward, character by character, from the incorrect choice. Select another path and try again.

5. Consider the following checklist of items to include in your notebook.

   a. Sketch the specimen, emphasizing its shape, and highlight the key characters used in the identification.

   b. Include notations about color, jaws and teeth, descriptions of fins, and characteristics of the family (cite the reference used).

6. Complete the required report for each species. Obtain any necessary details from reference textbooks.

Interpretation

1. Do the key characters correspond to characteristics of orders described in reference books? Try to explain any discrepancies.

2. For a selected species of fish, briefly outline its life history. Some aspects that might be discussed include habitat, range and migration, spawning cycle, diet and feeding habits, age, and growth. Identify the references used.

REFERENCES

Hoese, H. D., and R. H. Moore. 1977. *Fishes of the Gulf of Mexico, Texas, Louisiana, and adjacent waters.* Texas A&M Univ. Press, College Station, TX, 327 pp.

Murdy, E. O. 1983. *Saltwater fishes of Texas. A dichotomous key.* Publ. TAMU-SG-83-607. Texas A&M Sea Grant Program, College Station, TX, 220 pp.

Nelson, J. S. 1994. *Fishes of the world* (4th ed). Wiley-Interscience, New York, 600 pp.

Robins, C. R., R. M. Bailey, C. E. Bond, J. R. Brooker, E. A. Lachner, R. N. Lea, and W. B. Scott. 1980. *A list of common and scientific names of fishes from the United States and Canada* (4th ed). American Fisheries Society, Special Publication No. 12, 174 pp.

KEY TO ORDERS OF FISHES

This Key, which is modified from those of Hoese and Moore (1977) and Murdy (1983), is extensive in geographic coverage, but has its limitations. Space constraints prevent inclusion of all families belonging to the orders covered.

The families of fishes selected and listed under the orders occur over the continental shelf of the northwestern Gulf of Mexico, as well as temperate, subtropical, and tropical regions from Cape Hatteras, North Carolina, to Cape Canaveral, Florida, and from northwest Florida to southern Texas and Mexico. Many families occurring on the Pacific coast of North America are also listed in the Key.

The Key includes sharks, skates, rays, and bony fishes. Hagfishes, lampreys, and chimaeras are excluded.

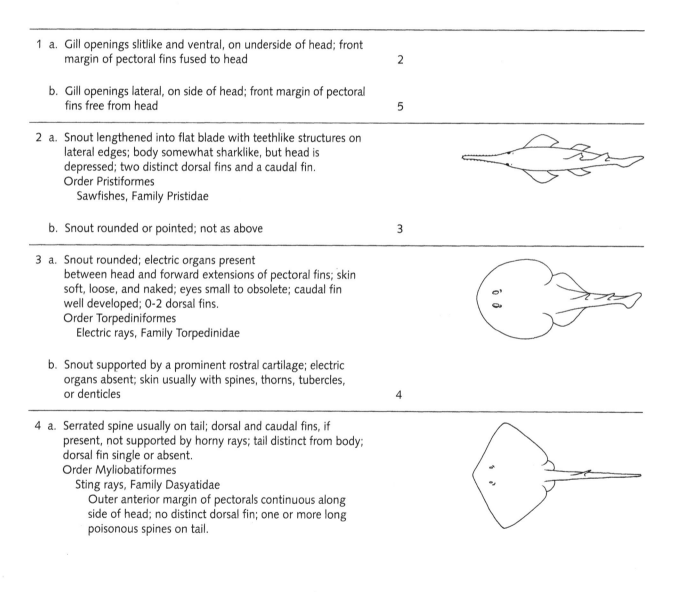

1 a.	Gill openings slitlike and ventral, on underside of head; front margin of pectoral fins fused to head	2
b.	Gill openings lateral, on side of head; front margin of pectoral fins free from head	5
2 a.	Snout lengthened into flat blade with teethlike structures on lateral edges; body somewhat sharklike, but head is depressed; two distinct dorsal fins and a caudal fin. Order Pristiformes Sawfishes, Family Pristidae	
b.	Snout rounded or pointed; not as above	3
3 a.	Snout rounded; electric organs present between head and forward extensions of pectoral fins; skin soft, loose, and naked; eyes small to obsolete; caudal fin well developed; 0-2 dorsal fins. Order Torpediniformes Electric rays, Family Torpedinidae	
b.	Snout supported by a prominent rostral cartilage; electric organs absent; skin usually with spines, thorns, tubercles, or denticles	4
4 a.	Serrated spine usually on tail; dorsal and caudal fins, if present, not supported by horny rays; tail distinct from body; dorsal fin single or absent. Order Myliobatiformes Sting rays, Family Dasyatidae Outer anterior margin of pectorals continuous along side of head; no distinct dorsal fin; one or more long poisonous spines on tail.	

Butterfly rays, Family Gymnuridae
  Disk extremely broad (more than 1.5 times wider than long); no caudal fin; tail short.

Eagle rays, Family Myliobatidae
  Head elevated and distinct from disk; eyes and spiracles lateral on head; gill openings about equal to length of eye; tail much longer than disk; venomous spines present in some; small dorsal fin; no caudal fin.

b. Tail lacks spine; dorsal and caudal fins well-developed and supported by horny rays; two dorsal fins.
  Order Rajiformes
    Skates, Family Rajidae
      Caudal fin reduced or absent; tail slender; 0–2 dorsal fins; most with prickles on skin.

Guitarfishes, Family Rhinobatidae
  Body intermediate between sharklike and skatelike; tail stout, not definitely marked off from body; two distinct dorsal fins and a caudal fin.

5 a.	Five to seven pairs of gill slits	6
b.	One pair of gill openings	16

6 a.	Six or seven gill slits; one dorsal fin. Order Hexanchiformes Cow sharks, Family Hexanchidae	
b.	Five gill slits; two dorsal fins	7

7 a.	Specialized nostrils connected to mouth by deep grooves; barbels usually present near front margin of nostrils; first dorsal fin posteriorly placed (above pelvics). Order Orectolobiformes Nurse sharks, Family Orectolobidae	
b.	Nasoral grooves absent; front margin of nostrils without barbels	8

8 a. Anal fin present   9

   b. No anal fin   15

9 a. Head flattened, with lateral expansions or lobes; eyes on
     lateral margins of lobes; spiracle absent.
     Order Carchariniformes (in part)
       Hammerhead sharks, Family Sphyrnidae

   b. Head tapered, no lateral expansions   10

10 a. Origin of first dorsal fin over or behind base of pelvic fin.
      Order Carchariniformes (in part)
        Cat sharks, Family Scyliorhinidae
          Nictitating membrane absent; fifth gill slit over origin of
          pectoral fin; first dorsal fin base longer than tail length
          and well anterior of pelvic fins.

   b. Origin of first dorsal fin well anterior to base of pelvic fin   11

11 a. Strong lateral keel on each side of tail; caudal fin lunate and
      nearly symmetrical; teeth large; fifth gill slit over origin of
      pectoral fin.
      Order Lamniformes (in part)
        Mackerel sharks, Family Lamnidae

   b. No keel on tail, or very weak; caudal fin asymmetrical, upper
     lobe longer than lower lobe   12

12 a. Caudal fin very long, length about half of total body length;
      third to fifth gill slits over origin of pectoral fin.
      Order Lamniformes (in part)
        Thresher sharks, Family Alopiidae

   b. Caudal fin length less than half total body length   13

13 a. Last gill slit well in front of origin of pectoral fin; eye without
      nictitating fold or membrane.
      Order Lamniformes (in part)
        Sand tigers, Family Odontaspidae

   b. Last gill slit over or behind origin of pectoral fin; eye with
     nictitating fold or membrane   14

14 a. Precaudal pit (dorsal and ventral indentations on tail in front
      of caudal fin) present; nictitating eyelid present; spiracle, if
      present, a narrow slit; teeth bladelike, with one cusp.
      Order Carcharhiniformes (in part)
        Requiem sharks, Family Carcharhinidae

   b. Precaudal pits absent; nictitating eyelid visible only in front
     and back corners of eye; spiracle oval; teeth small, low,
     rounded with three cusps.
     Order Carcharhiniformes (in part)
       Smooth dogfish sharks, Family Triakidae

15 a. Eyes lateral; trunk nearly rounded in cross section; front margin of pectoral fin does not overlap gill slits; each dorsal fin preceded by a spine.
   Order Squaliformes
      Dogfish sharks, Family Squalidae

   b. Eyes dorsal, trunk dorso-ventrally flattened; front margin of enlarged pectoral fins overlap gill slits; two spineless dorsal fins; no anal fin; five gill slits; spiracle large; mouth almost terminal with barbels on anterior margin.
   Order Squatiniformes
      Angel sharks, Family Squatinidae

16 a. Caudal fin abbreviate heterocercal   17

   b. Caudal fin present or absent but not as above   18

17 a. Body covered with cycloid scales (bowfins) or skin embedded with bony plates (sturgeons) (both freshwater species)

   b. Body covered with ganoid scales; dorsal fin far back, with few rays; length of base of dorsal fin less than half total body length; snout beaklike; mouth with needlelike teeth; abbreviated heterocercal tail (freshwater with tolerance for saltwater).
   Order Lepisosteiformes
      Gars, Family Lepisosteidae

18 a. Snout a protruding tube with short jaws at end, or if not protruding, upper jaw protractile.
   Order Syngnathiformes
      a. Pipefishes and b. seahorses, Family Syngnathidae
         Body elongated and encased in series of bony rings; one dorsal fin; anal fin very small; no pelvics.

a

b

      Cornet fishes, Family Fistulariidae
         Body depressed, elongated, and naked, or with minute prickles and linear series of scutes (no scales); no barbel on jaw; no dorsal spines; caudal fin forked with elongate filament produced by middle two caudal rays; lateral line well developed, arched anteriorly almost to middle of back.

Trumpet fishes, Family Aulostomidae
    Body compressed, elongated, and scaly; fleshy barbel at
    tip of lower jaw; caudal fin rounded; lateral line well
    developed; anus far behind pelvics.

Snipefishes, Family Macrorhamphosidae
    Body compressed, deep, usually with bony plates on
    each side of back; no barbel on jaw; lateral line present
    or absent.

Order Gasterosteiformes
    Sticklebacks, Family Gasterosteidae
        Body elongated or not, with lateral bony scutes or naked;
        series of 3–16 well-developed isolated dorsal spines
        followed by a normal dorsal fin; pelvic fin with one spine
        and one or two soft rays.

b. Snout not protruding as tube, but may form beak whose jaws
   extend to its base   19

---

19 a. Gill opening a small hole behind base of pectoral fin; anterior
      dorsal fin rays modified as angling device with lure,
      sometimes retracted under snout.
      Order Lophiiformes
        Goosefishes, Family Lophiidae
            Huge, wide, flattened head; teeth well developed; fringe
            of small flaps extending around lower jaw and along
            sides of head onto the body.

        Frogfishes, Family Antennariidae
            Body covered with loose skin, naked or with denticles;
            first three spines separate; gill opening below base of
            pectoral.

        Batfishes, Family Ogcocephalidae
            Body depressed and flattened ventrally; relatively short
            fishing lure; mouth nearly horizontal; gill opening in or
            above pectoral base.

b. Gill opening anterior to base of pectoral fin; no angling device   20
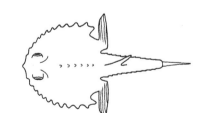

---

20 a. Body asymmetrical and flat, with both eyes on one side of head.
    Order Pleuronectiformes
        Right-eye flounders, Family Pleuronectidae
            Eyes on right side; origin of dorsal fin above the eyes; lateral line well developed on both sides; pelvic fins symmetrical.

        Left-eye flounders, Family Bothidae
            Eyes on left side; pelvic fin bases short and nearly symmetrical; pelvic fins without a spine; pectoral and pelvic rays branched.

        Soles, Family Soleidae
            Eyes (small) on right side and close together; dorsal and anal fins free from caudal fin or united with caudal; pectorals present.

        Tonguefish, Family Cynoglossidae
            Eyes on left side; dorsal and anal fins confluent with pointed caudal fin; usually only left pelvic fin developed; pectorals absent; eyes very small and close together; mouth asymmetrical.

    b. Body symmetrical, one eye on each side of head   21

21 a. Top of head with flat, oval, sucking disk with transverse grooves.
    Order Perciformes (in part)
        Remoras, Family Echeneidae
            Body elongated, head flattened, lower jaw projecting past upper jaw; scales small, cycloid; dorsal and anal fins lacking spines.

    b. Top of head lacks sucking disk   22

22 a. Breast with large sucking disk (modified pelvics).
    Order Scorpaeniformes (in part)
        Lumpfishes, Family Cyclopteridae
            Body globose, usually covered with tubercles.

    Order Gobiesociformes
        Clingfishes, Family Gobiesocidae
            Each pelvic with one small spine and four soft rays; single dorsal fin without spines.

    b. Breast without sucking disk   23

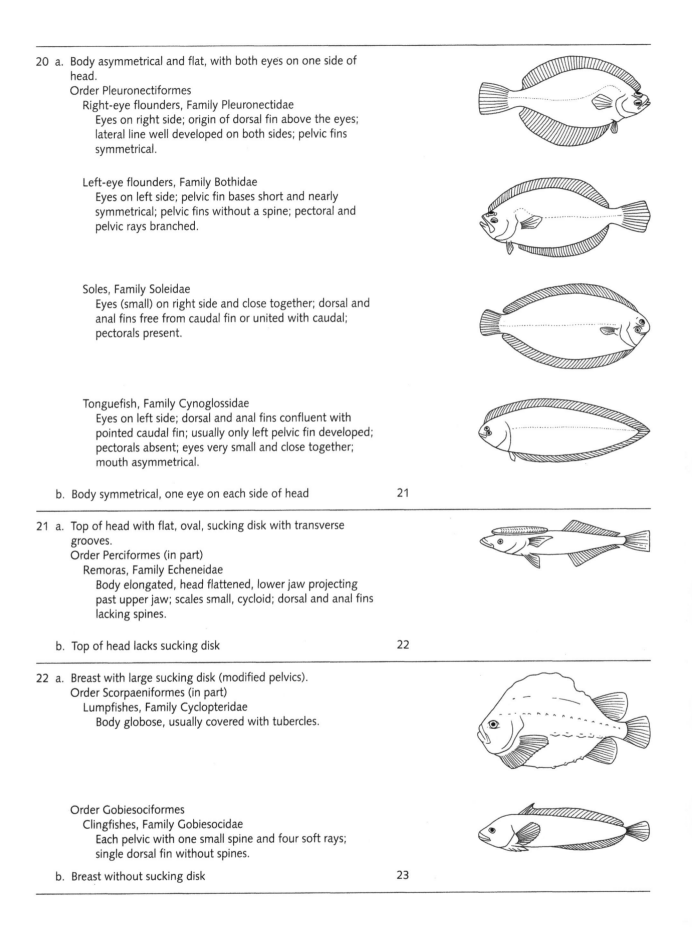

23 a. Lateral line single, located below longitudinal midline along
       entire length.
       Order Beloniformes
          Needlefishes, Family Belonidae
             Mouth opening large; both jaws elongated with
             numerous needlelike teeth.

          Halfbeaks, Family Hemirhamphidae
             Upper jaw much shorter than lower; pectoral and pelvic
             fins short.

          Flying fishes, Family Exocoetidae
             Jaws relatively short and equal in length; exceptionally
             large pectoral fins; pelvics exceptionally large in some
             species; lower lobe of caudal fin longer than dorsal lobe.

          Sauries, Family Scomberesocidae
             Mouth opening relatively small; jaws varying in length
             from both produced into long, slender beaks to relatively
             short beaks with lower jaw only slightly produced.

    b. Lateral line present or absent; if present, single or multiple and
       located at least partly along or above longitudinal midline   24

24 a. Gill opening earlike hole or slit anterior to, or slightly above
       base of, pectoral fin, opening seldom longer than width of
       base of pectoral; body never eel-like.
       Order Tetraodontiformes
          Puffers, Family Tetraodontidae
             Body naked or with only short prickles; four fused teeth
             in jaws; caudal fin moderately forked to rounded.

          Porcupine fish, Family Diodontidae
             Body covered with well-developed sharp spines; two
             fused teeth in jaws.

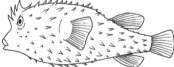

          Filefishes, Triggerfishes, Family Balistidae
             Body usually compressed; head and body usually
             covered with scales; no pelvic fins; first dorsal spine with
             locking mechanism; upper jaw not protractile, with two
             rows of protruding incisorlike teeth.

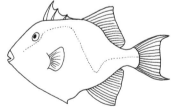

          Trunkfishes, Family Ostracidae
             Body encased in bony carapace; no pelvic fins.

    b. Gill opening not as above, except in eel-like forms   25

25 a. Both pectoral and dorsal fins with single strong front spine (hard ray); head barbels well developed and elongate; body naked; adipose fin present.
   Order Siluriformes
      Catfishes, Family Ariidae
         Caudal fin forked; usually three pairs of barbels; pectoral and dorsal fins with a spine.

   b. If present, spines of pectoral and dorsal fins not in above combination; head barbels present or absent; body naked or with scales; adipose fin present or absent   26

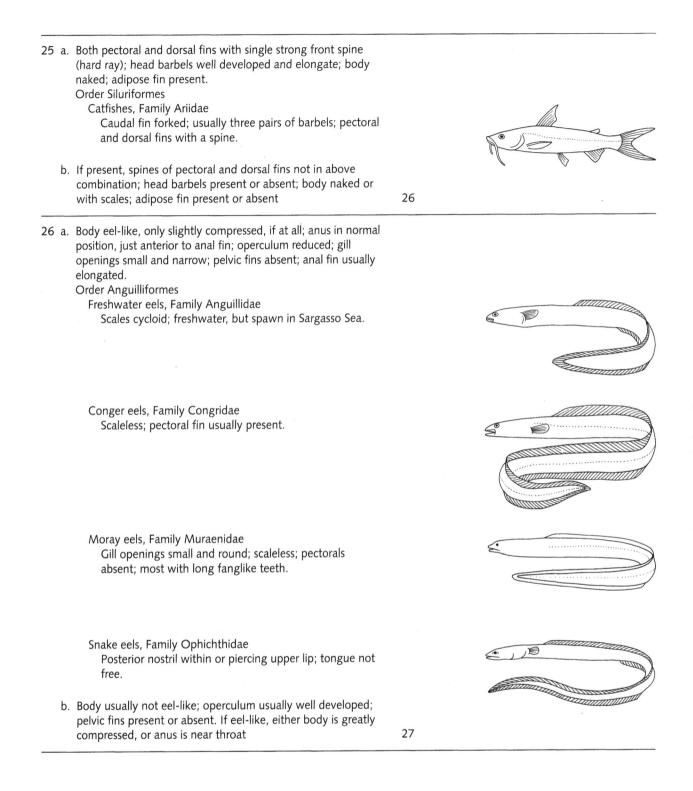

26 a. Body eel-like, only slightly compressed, if at all; anus in normal position, just anterior to anal fin; operculum reduced; gill openings small and narrow; pelvic fins absent; anal fin usually elongated.
   Order Anguilliformes
      Freshwater eels, Family Anguillidae
         Scales cycloid; freshwater, but spawn in Sargasso Sea.

      Conger eels, Family Congridae
         Scaleless; pectoral fin usually present.

      Moray eels, Family Muraenidae
         Gill openings small and round; scaleless; pectorals absent; most with long fanglike teeth.

      Snake eels, Family Ophichthidae
         Posterior nostril within or piercing upper lip; tongue not free.

   b. Body usually not eel-like; operculum usually well developed; pelvic fins present or absent. If eel-like, either body is greatly compressed, or anus is near throat   27

27 a. Gular plate (large bone in throat between angle of lower jaws) present; gill openings wide; caudal fin deeply forked; scales cycloid.
Order Elopiformes

    Tarpons, Family Megalopidae
      Body compressed; mouth terminal or superior; insertion of pelvic fin beneath or in advance of origin of dorsal.

    Ladyfish, or tenpounders, Family Elopidae
      Body rounded (little compressed); mouth terminal; insertion of pelvic fin beneath or posterior to origin of dorsal fin.

    Bonefish, Family Albulidae
      Insertion of pelvic fin near posterior end of dorsal.

  b. Gular plate absent   28

28 a. Eye with crescent of white tissue over upper part of iris.
Order Salmoniformes

    Argentines, Family Argentinidae
      Body trunk elongated and angular in cross section; adipose fin over anal fin base; dorsal fin origin in front of pelvics; pectoral fin base on ventrolateral surface; mouth small.

    Smelts, Family Osmeridae
      Adipose fin usually present; forked tail.

    Salmon, Family Salmonidae
      Scales small, more than 110 along lateral line; caudal fin emarginate or truncate.

  b. Eye not as above; trunk of body not as above   29

29 a. Upper jaw formed into bony, swordlike bill.
Order Perciformes (in part)

    Swordfish, Family Xiphiidae
      Scales absent in adult; pelvic fins absent; caudal peduncle with single median keel on each side.

    Billfishes, Family Istiophoridae
      Scales present; pelvic fins elongated; caudal peduncle with two keels on each side; dorsal fin with very long base, depressible into groove.

  b. Upper jaw not as above   30

30 a. Pelvic fins, when present, without spines; when pelvic fins absent, anus near throat, and body trunk nearly cylindrical (eel-like)   31

     b. Pelvic fins, when present, with spines; when pelvic fins absent, anus in normal position, and body trunk usually compressed   36

31 a. Pelvic fins present or absent; when present, inserted directly under, or anterior to, pectoral fin; if slightly behind pectoral base, then body tapers to a point posteriorly   32

     b. Pelvic fins present and inserted behind base of pectoral fin; if only slightly behind, then body does not taper to a point   33

32 a. Each pelvic fin with one or two soft rays, or pelvic fins absent; pelvic fins inserted near level of preopercle (forepart of operculum) or farther anterior; dorsal and anal fins with long bases, extend to and often join with caudal fin.
     Order Ophidiiformes
         Cusk eels, Family Ophidiidae
             Dorsal fin rays usually equal to or longer than opposing anal fin rays; anus and anal fin origin usually behind tip of pectoral fin; scales present.
         Pearlfishes, Family Carapidae
             Anal fin rays longer than opposing dorsal fin rays; anus of adults and anal fin origin far forward, behind head and usually beneath pectoral fin; scales absent.

     b. Caudal fin separate from dorsal and anal fins, only rarely connected (in macrourids); gill openings wide, extend above base of pectoral fin.
     Order Gadiformes
         Codlets, Family Bregmacerotidae
             Two dorsal fins and one long anal fin (first dorsal fin on nape and consisting of one elongated ray, second dorsal and anal fins with large notch in middle); no chin barbel.

         Hakes, Family Merlucciidae
             Two dorsal fins and one anal; no chin barbel; caudal fin truncate, not confluent with dorsal and anal fins.

         Grenadiers, Family Macrouridae
             Long tapering tail; dorsal and anal fins confluent with caudal fin or with each other.

         Codfishes, Family Gadidae
             First dorsal posterior to head; three dorsal fins and two anal fins; chin barbel usually present; caudal fin truncate or slightly forked.

33 a. Adipose fin or detached finlet present   34

   b. Adipose fin or finlet absent   35

34 a. Numerous photophores (light-producing organs) present on
      ventral aspect of body.
      Order Myctophiformes
         Lantern fishes, Family Myctophidae
            Small photophores in groups and rows on head and
            body; scales usually cycloid (deep water species, but
            migrates to between 10 m and 100 m depth).

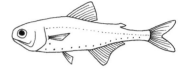

   b. Body lacks photophores.
      Order Aulopiformes
         Lizardfishes, Family Synodontidae
            Adipose fin usually present.

         Greeneyes, Family Chlorophthalmidae
            Eyes large, normal; tip of upper jaw not extending
            beyond orbit of eye.

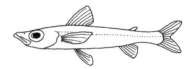

         Lancetfishes, Family Alepisauridae
            Body slender, covered with pores; scale absent; dorsal
            fin high and extending along most of body.

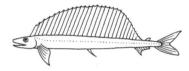

35 a. Caudal fin forked; adipose eyelid usually present; scales
      lacking on head and operculum.
      Order Clupeiformes
         Herring, Family Clupeidae
            Head scaleless; dorsal and pelvic fins rarely absent;
            mouth inferior, superior, or terminal; teeth small or
            absent; lateral line not extending onto body.

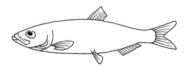

         Anchovies, Family Engraulidae
            Tip of snout overhangs mouth; upper jaw extends well
            beyond eye.

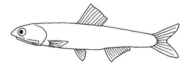

   b. Caudal fin rounded or truncate; adipose eyelid absent; scales
      on head and/or operculum.
      Order Cyprinodontiformes
         Killifishes, Family Cyprinodontidae
            Lateral line chiefly on head; pelvic fin bases relatively far
            apart.

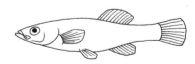

36 a. Three gill-bearing arches; dorsal fin divided into two parts: spinous part short with two to three low stout spines, soft part long with many segmented rays.
        Order Batrachiformes
            Toadfishes, Family Batrachoididae
                Three dorsal spines and opercular spine; body scaleless or with cycloid scales; usually one to three lateral lines.

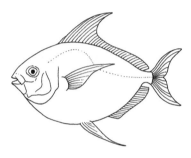

    b. Four gill-bearing arches; dorsal fin continuous or divided, variable lengths   37

37 a. Body deep and compressed, with anal fin in two parts: first with three stout spines connected by membrane, second part with 24–33 soft rays; *or* body deep and compressed with no fin spines, and pectoral fins with horizontal bases   38

    b. Not as above   39

38 a. Pectoral fins with horizontal bases; no spines in fins.
        Order Lampriformes
            Opahs, Family Lampridae
                Body oval and compressed; lateral line arched high in front; dorsal and anal fins long; minute cycloid scales.

    b. Pectoral fins with near vertical bases; fins with stout spines.
        Order Zeiformes
            Dories, Family Zeidae
                Small spines at base of dorsal and anal fin rays; eight or nine spinous plates along abdomen; scales small, rudimentary, or absent.

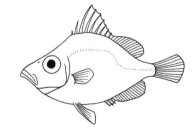

            Boarfishes, Family Caproidae
                Body covered with small ctenoid scales; no abdominal spinous plates; caudal fin rounded.

39 a. Pelvic fins present, with one spine and six to ten rays; eyes usually large.
Order Beryciformes

Beardfishes, Family Polymixiidae
Body moderately elongated and compressed; dorsal fin continuous; pelvic fins subabdominal.

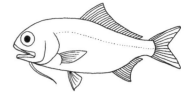

Squirrelfishes, Family Holocentridae
Long dorsal fin with spiny portion and soft-rayed portion divided by a notch; caudal fin forked; scales large and ctenoid; opercle with spiny edge, strong spine present at angle; longest anal spine usually longer than or equal to longest dorsal spine.

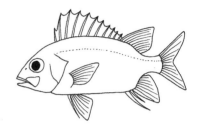

b. Pelvic fins present or absent; if present, with one spine and not more than five rays   40

40 a. Pectoral fins high on sides; dorsal fin divided into two well-separated parts, spinous dorsal with four to eight slender spines or unsegmented hard rays; anal fin with one weak spine or unsegmented ray; lateral line absent; sides with prominent dark or silvery longitudinal band.
Order Atheriniformes

Silversides, Family Atherinidae
Pelvic fins abdominal; scales relatively large (31–50 in lateral series).

b. Not as above   41

41 a. Suborbital bony stay (spinous bony process) extends across cheek to preoperculum; head and body spiny or bony-plated.
Order Scorpaeniformes (in part)

Mail-cheek fishes, Scorpion fishes, Family Scorpaenidae
Body compressed; head usually with ridges and spines; scales, when present, usually ctenoid; dorsal fin usually single, with a notch; venom gland in dorsal, anal, and pelvic spines.

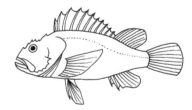

Sea robins, Family Triglidae
Two separate dorsals; bony head; lower two or three pectoral rays enlarged and free; body with scales or covered by plates.

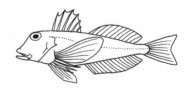

b. Lacks suborbital bony stay.
Order Perciformes. No single set of characteristics unites the members of this incredibly large and diverse group of fishes. Brief descriptions of some of the more common families follow.

Sea basses, Serranidae
Opercle with three spines; scales usually ctenoid, cycloid in some; lateral line complete; dorsal fin generally continuous, may be notched; three anal fin spines; caudal fin usually rounded, truncate, or lunate.

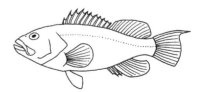

Jacks, Carangidae
Body generally compressed; small cycloid scales in most species, ctenoid in a few; up to nine detached finlets sometimes present behind dorsal and anal fins; two dorsal fins in adults; caudal fin widely forked; caudal peduncle slender.

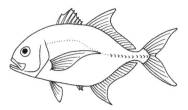

Snappers, Lutjanidae
Dorsal fin continuous or with shallow notch; anal fin with three spines; pelvics inserted behind pectoral base; mouth terminal, moderate to large; most with enlarged canine teeth on jaws.

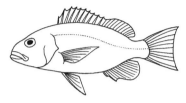

Grunts, Haemulidae (Pomadasyidae)
Dorsal fin continuous; anal fin with three spines; mouth small; enlarged chin pores usually present.

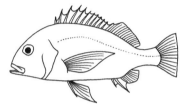

Porgies, Sparidae
Dorsal continuous; anal with three spines.

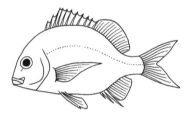

Drums, Sciaenidae
Dorsal long, with deep notch separating spinous from soft portion; anal with one or two spines; lateral line extending to end of caudal fin; caudal slightly emarginate to rounded; single barbel or patch of small barbels on chin of some species.

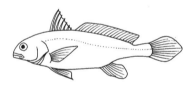

Goatfishes, Mullidae
Body elongated; two widely separated dorsal fins; anal with one or two small spines; two long chin barbels; caudal forked.

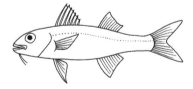

Butterflyfishes, Chaetodontidae
Body strongly compressed; no spine at angle of preopercle; dorsal continuous or with slight notch; caudal rounded or emarginate; mouth small, terminal, protractile.

Angel fishes, Pomacanthidae
  Body strongly compressed; strong spine at angle of
  preopercle; dorsal and anal fins with elongate extension
  on hind margin in many species.

Damselfishes, Pomacentridae
  Body usually high and compressed; mouth small; lateral
  line incomplete or interrupted; anal fin with two spines;
  single continuous dorsal fin.

Mullets, Mugilidae
  Widely separated spiny-rayed and soft-rayed dorsal fins;
  pelvics subabdominal; lateral line absent or very faint;
  mouth moderate in size; teeth small or absent; gill
  rakers long.

Barracuda, Sphyraenidae
  Body elongated; mouth large, jutting lower jaw with
  strong fanglike teeth; lateral line well developed; two
  widely separated dorsal fins.

Wrasses, Labridae
  Mouth protractile; jaw teeth mostly separate, usually
  projecting outward; lateral line continuous or
  interrupted; scales cycloid, generally large to moderate.

Parrot fishes, Scaridae
  Mouth nonprotractile; jaw teeth coalesced; scales large
  and cycloid.

Eelpouts, Zoarcidae
  Body elongated; dorsal and anal fins long and confluent
  with caudal fin; mouth subterminal.

Gunnels, Pholidae
  Dorsal fin about twice as long as anal; pectoral fins
  small, rudimentary, or absent; lateral line short or absent.

Blennies, Blenniidae
  Body naked or with modified scales; head blunt; pelvics
  anterior to pectorals.

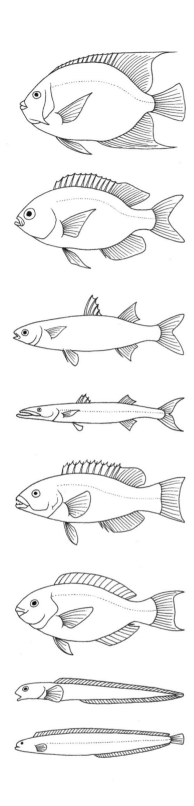

Sand lances, Ammodytidae
Scales cycloid, minute; pelvics usually absent; lateral line high, close to dorsal fin; no teeth; single long dorsal fin.

Gobies, Gobiidae
Pelvic fins united, usually forming an adhesive or sucking disk; spinous dorsal, when present, separate from soft dorsal; scales cycloid or ctenoid.

Surgeon fishes, Acanthuridae
Body compressed; one or more spines on caudal peduncle (which, when extended, can form a weapon).

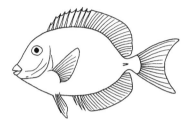

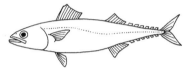

Mackerels, Scombridae
Two dorsal fins (depressible into grooves) with finlets behind second dorsal and anal fins; scales cycloid and small; slender caudal peduncle with two keels; first dorsal fin origin well behind head; pectorals high on body; pelvics beneath pectorals.

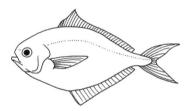

Butterfishes, Stromateidae
Body deep; pelvic fins absent in adult; dorsal continuous.

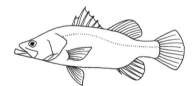

Snooks, Centropomidae
Lateral line extending onto tail; dorsal fin in two portions; anal fin with three spines; caudal rounded, truncate, or forked.

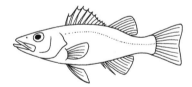

Temperate basses, Percicthyidae
Opercle with two rounded spines; lateral line complete; caudal usually forked.

Internal Anatomy of a Bony Fish

OBJECTIVES

After completing this unit, you will be able to

- Understand the internal anatomy of marine bony fish and relate structure to adaptation; and

- Appreciate host–parasite interactions.

INTRODUCTION

As mentioned in Unit 9, bony fishes are a highly diverse group, which is reflected in their morphology. This unit introduces the internal anatomy of bony fishes. Dissection of fresh specimens allows you to observe tissues and organs in their natural condition, and also presents an opportunity to explore the realm of fish parasites. These organisms, at some stage of their lives, require some vital factor that can be obtained from other living organisms—their hosts.

Nearly every species of fish harbors one or more parasites during some part of its life. Individual fish may have several parasite species living on or in the body, in incredibly large numbers. Fish commonly shelter hundreds of one species on the gills, and similar numbers of other species in the gut. Few fish parasites are pathogenic, but some—especially protistans—can cause extensive harm and reach epidemic proportions.

The following description applies to a typical bony fish such as a perch or a bass. It is important to recognize that variations from this stereotype are related to the kinds and degrees of adaptations that have evolved among these animals.

GENERAL PROCEDURE

1. Perform the dissections as directed. Refer to the accompanying figures, and record your observations in your notebook as you work.
   NOTE: The specimens provided may be freshly acquired from a local seafood vendor. If fresh specimens are unavailable, preserved specimens will be used.

2. If more than one species is available for observation, compare them as you proceed through this exercise. Your answers on the laboratory report should then be written with a comparative viewpoint.

3. Refer to aspects of external morphology (Unit 9) as necessary.

4. Discard biological waste, including the carcass, in the appropriate containers. Waste from fresh specimens must never be mixed with preserved waste. Waste toweling must not be mixed with biological waste.

5. Look for parasites as you perform the internal dissection. Parasites will be discussed in the appropriate sections of the procedures. If you discover any, transfer them to a slide or small dish with seawater, and notify your instructor. Most will require examination with a compound microscope.

EXERCISE 1
Dissecting a Typical Bony Fish

Skeletal System

PROCEDURE

Examine mounted fish skeletons on display.

1. The **skull** (Fig. 10.1a) has two distinct parts: (1) the **neurocranium** forms the floor and roof to the brain case, surrounds and protects the olfactory (smell), optic (sight), and otic (hearing) or-

gans, and gives form to the face; and (2) the **branchiocranium** includes the mandibles, hyoid arch (supports tongue), operculum, and branchial (gill) arches.

2. **Vertebrae** (Fig. 10.1b,c) of fishes are less complex than those of land vertebrates. Articulation takes place by simple contact of the **centra.** Throughout the length of the **vertebral column,** a series of dorsal **neural arches** (with **neural spines**) protects the spinal cord that lies in the **neural canal.** Ventral to each centrum in the tail is a **hemal arch** (with **hemal spines**) that partially encases the axial blood vessels that lie in the **hemal canal. Dorsal ribs (intermuscular bones)** are embedded in the trunk musculature.

Ventral ribs develop within the connective tissue partitions **(myosepta)** of the lateral muscle bundles **(myotomes).**

3. **Fin rays** of the **caudal fin** are supported by modified vertebrae in the **caudal peduncle** (Fig. 10.2a). In the **dorsal** and **anal fins,** the **fin rays** are supported by a series of small bones, called **pterygiophores,** embedded in the median musculature (Fig. 10.2b).

4. The **pectoral girdle** (Fig. 10.2c) consists of numerous bony and cartilaginous elements. The rays of the fins connect to the girdle by small skeletal elements.

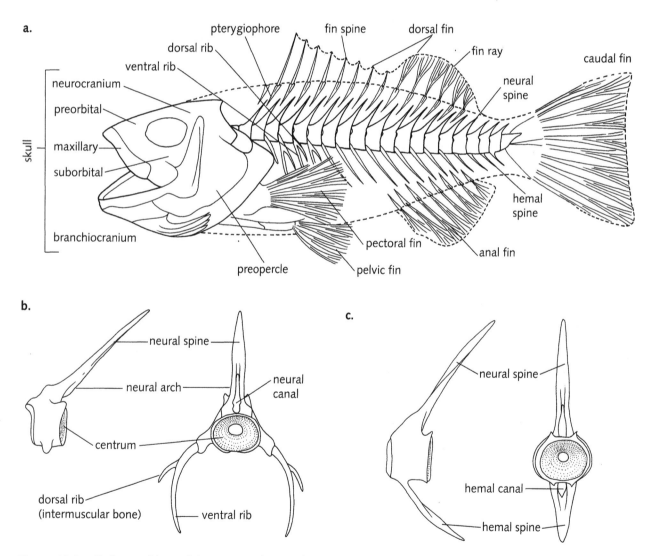

Figure 10.1 Skeleton of bony fish. (a) Lateral view of entire, articulated skeleton. (b) Abdominal vertebrae. (c) Caudal vertebrae.

5. The **pelvic girdle** consists of a single pair of bones that may be separate or fused, the condition shown in Figure 10.2d. Articulation with the fin rays varies from species to species. The girdle is abdominal in position in more primitive (evolutionarily) fishes, such as herring and salmon. In more advanced bony fishes, such as bass, the girdle is in close contact with the pectoral fins and thoracic in position (see Fig. 9.4 in Unit 9). Where are the paired fins located in the mounted skeleton? How would you describe the relative evolutionary position of your specimen?

Musculature

PROCEDURE

1. Skin a portion of one side of the body to expose the trunk musculature (Fig. 10.3a). Slice the skin vertically behind the head and horizontally along, and just beneath, the dorsal fin. Pull the skin away carefully with your fingers as you clear the musculature from the skin with forceps or scalpel. Note the shape of the lateral muscle bundles **(myotomes)**. They resemble Ws that are turned on their sides and stacked together. A layer of connective tissue **(horizontal septum)** divides the myotomes into dorsal **epaxial muscles** and ventral **hypaxial muscles**. Caudally, both sets participate in locomotion. Cranially, the hypaxial muscles support the body viscera. *PARASITE WATCH:* Nematodes (round worms) are commonly encysted in muscle tissue. Look for opaque nodules embedded in the tissue. Cysts may be sliced open to reveal the worms.

2. Try to separate the myotomes. Observe the direction of the muscle fibers. You might need to use a dissecting microscope for this examination. Do they run zigzag like the myotomes, or

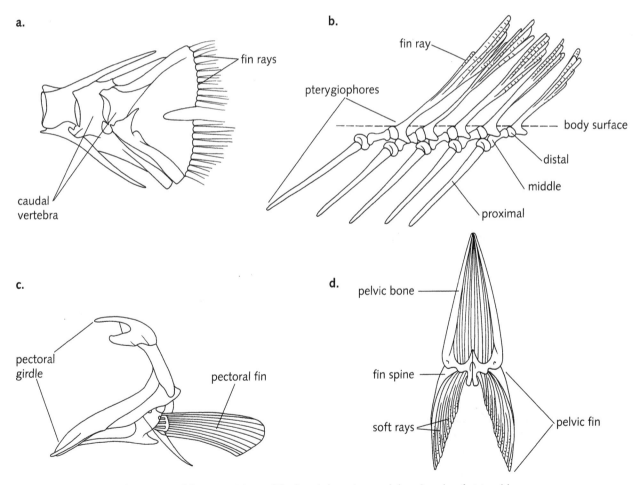

Figure 10.2 Skeletal support of fin rays. (a) Modified vertebrae in caudal peduncle of striped bass. (b) Pterygiophores of median (dorsal) fin. (c) Lateral view of left pectoral bones and fin. (d) Ventral view of pelvic skeleton.

are they parallel to the long axis of the body? Try to trace single fibers from the superficial origin on the body surface through the muscle mass to the deeper insertion. Locating the exact point is nearly impossible.

3. Is all musculature white or is some darkened? Most bony fish have a midlateral muscle mass that is shallow and reddish in color (Fig. 10.3b). In some fish, such as the tuna and swordfish, the red muscle mass is disproportionately large. The red color results from the presence of **myoglobin** and enhanced blood supply to the muscle. What is myoglobin's function? How might the existence of this muscle mass enhance sustained swimming ability?

4. Make a cross section of your specimen posterior to the anus. Compare what you see with Figure 10.3b. Sketch the cross section.

5. If the aquarium contains live fishes, watch their swimming motions and try to observe the use of body muscles and fins in locomotion and in maintaining position. Which are used for propulsion? Which for stability? Are their movements coordinated? Does fin use vary with body form?

Mouth Cavity, Pharynx, and Branchial System

PROCEDURE

1. Lift, or cut away, the operculum from the left side to expose the **gill arches** (Fig. 10.4a). Count them. Does your specimen have a **pseudobranch**—a group of gill filaments on the inner wall of the operculum? Sketch a gill arch *in situ* (in the fish). *PARASITE WATCH:* This region of the fish often holds ectoparasites, such as spiny-headed worms (phylum Acanthocephala), parasitic copepods, fish lice, and trematode flatworms. They may be attached to, or embedded in, the walls of the chambers, or in the gill filaments. Remove them by cutting around their point of attachment with a scalpel.

2. Remove one arch, place it in water, and examine it with a dissecting microscope. Note the double row of **gill filaments** on the posterior, or aboral, side of the arch (Fig. 10.4b,c). They are filled with blood capillaries, and are bright red in fresh specimens. Make a cross section of the arch, and examine the cut end under magnification. Locate the major blood vessels. Draw the details of a gill arch.

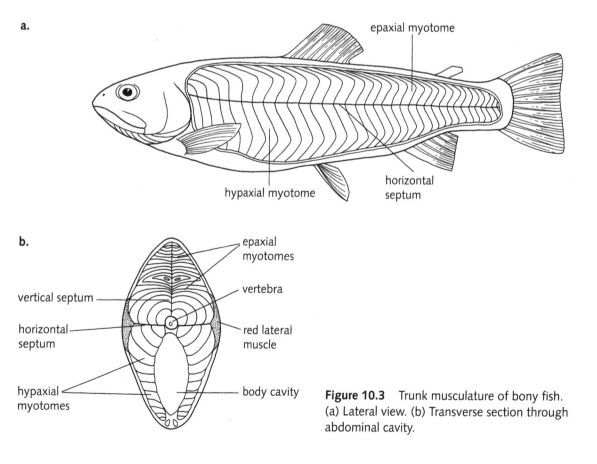

Figure 10.3 Trunk musculature of bony fish. (a) Lateral view. (b) Transverse section through abdominal cavity.

3. Extending anteriorly from the gill arches are the **rakers** (Fig. 10.4b). They protect the filaments from abrasion by ingested coarse materials. The rakers' form is related to food and feeding habits of the species (Fig. 10.4d). They are stubby, rigid, and unadorned in omnivores and predators, but elongated, numerous, more flexible, and variously ornamented in plankton feeders. Describe the rakers in your specimen. Can you relate raker form in your specimen to diet? Does your conclusion agree with the dentition of your specimen (see Unit 9)?

4. Cut through the angle of the left jaw to expose the spacious **pharynx.** Note the arrangement and size of the gill arches, and the presence or absence of **pharyngeal teeth** (see Fig. 9.10 in Unit 9). These teeth can be found on both the roof and floor of the pharynx; they hold or grind prey prior to swallowing.

5. Observe any live fishes in the aquarium and attempt to understand the sequence of events involved in ventilation. The mechanics of water movement involve a continuous pumping action in which, by muscular action, the gill arches are pushed out laterally while the opercula are pressed against the body. This activity simultaneously enlarges the branchial cavity and closes its exit so that water flows into the mouth and pharynx. The branchial chambers are then compressed, the flaplike **oral valves** close, the gill covers open, and the water is forced out over the gills.

Abdominal Cavity

NOTE: Depending upon the degree of freshness of your specimen, it may have been necessary to inject the abdominal cavity with a preservative. Your instructor will notify you if that has occurred.

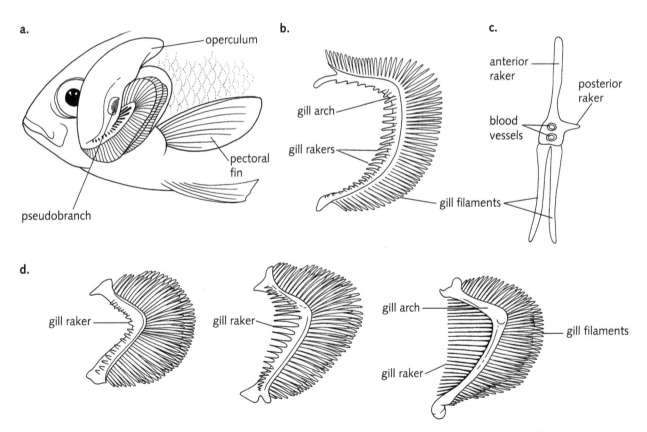

Figure 10.4 Arrangement and structure of gill arches of a bony fish. (a) Position of arches in left gill cavity, operculum lifted; pseudobranch attached to inner wall of operculum. (b) Lateral view of one arch. (c) Cross section through the gill arch. (d) Relationship of gill rakers to feeding habits, showing progressively efficient sieve potential from left to right.

The internal organs of the fish are analogous in function to those of mammalian systems.

PROCEDURE

1. Insert a point of the scissors near the anus. Being careful not to damage internal organs, cut anteriorly on the left side of the midventral line to a region anterior to the pelvic fins (or below the pectoral fins, whichever is more cranial) (Fig. 10.5). Extend your cut in the body wall dorsally from the anus until you contact the body musculature. Make another cut dorsally behind the pectoral fin, and remove the body wall by cutting between these two incisions. Note the shiny lining **(peritoneum)** on the body wall. Is it pigmented? What color is it?

2. The **intestine** is conspicuous because it is frequently encased in yellow fat. Remove enough of the fat to trace the digestive tract anteriorly. The intestine is relatively short in carnivores, but is often elongated and arranged in many folds in predominantly herbivorous species. Does the nature of this intestine correspond with your prior decisions on feeding habits of this species?

PARASITE WATCH: Look for parasitic flatworms, both trematodes and tape worms, in this region. The former occur in nearly every organ; the latter are usually found in the intestine. Slice open the intestine and flush the contents into a dish of seawater. Examine the contents of the dish, as well as the inner lining of the intestine. Because tape worms are usually embedded by their heads, they will need to be cut from the intestinal wall.

3. Find the **stomach,** which lies dorsal and somewhat to the left of the intestine. In carnivorous fishes, the stomach is elongated; in omnivores, it is often sac-shaped. The stomach wall can also be modified by muscular thickening and used as a grinding organ. Some fishes do not have a stomach, however. Describe the stomach in your specimen.

4. You may not be able to see the **esophagus,** but you can run a blunt probe through the mouth and into the opening of the esophagus in the pharynx. A highly expandable organ, the esophagus facilitates the swallowing of large prey.

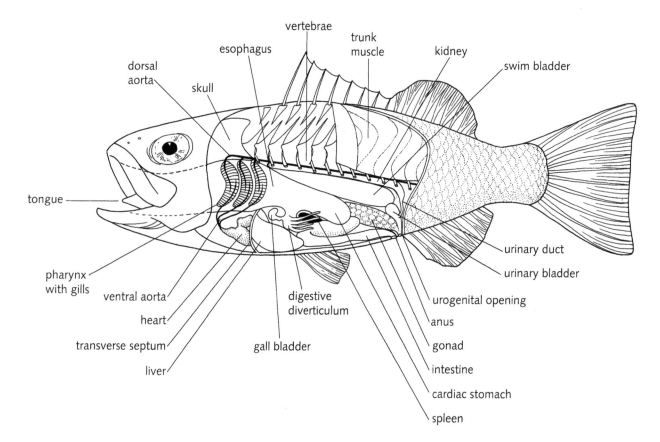

Figure 10.5   Lateral view of internal anatomy of a bony fish. Operculum and left body wall removed.

5. Anterior and ventral to the stomach is the **liver,** which is relatively large, lobed, and dark red in live specimens. A **gall bladder** is located under the right lobe. Lift the liver and locate the **cardiac portion** (a blind pouch) of the stomach. The stomach empties into the **duodenum,** the first part of the small intestine. Digestive diverticula (three or more fingerlike extensions of the gut) extend from the end of the duodenum nearest the stomach. Follow the intestine to the anus, noting the supply of blood vessels in the supporting membrane **(mesentery).**

6. The **spleen** is a dark, slender organ located between the stomach and the intestine.

7. The **pancreas,** a diffuse and indistinct organ, either lies in the fold of the duodenum or is embedded in the liver.

8. Open the stomach and empty its contents into a glass bowl. Examine the contents and compare them with other students' findings. Do the contents verify or refute your earlier diagnosis of food and feeding in this species? Explain. Do you see any parasites?

9. The **swim bladder,** if present, is a long, shiny sac that fills most of the body cavity dorsal to the visceral organs. It may be thin- or thick-walled. In some fishes, it connects with the alimentary canal; in others, it does not. Cut a slit in it and observe the internal structure. The bladder is a hydrostatic organ that adjusts the specific gravity of the fish to varying depths of water. Does your species have a swim bladder? What type is it? Observe whether bands of muscle appear within the bladder wall, or whether any originate on the dorsal body wall and insert on the bladder. Can you deduce the purpose of that musculature?

10. Observe fishes in the aquarium and determine which have swim bladders. Is the degree of buoyancy related to maneuverability of the fish? Are body shape, fin shape, and buoyancy correlated?

11. Most marine fishes are dioecious; individuals are either male or female. It is difficult to distinguish their sex externally.

    a. The single **ovary** lies caudal to the stomach, just ventral to the swim bladder (if present) and dorsal to the intestine. The size of the ovary varies seasonally, being largest just prior to spawning. An extension of the ovary serves as an **oviduct** that carries eggs to the **urogenital** pore just posterior to the anus.

    b. In the male, two elongated **testes** lie dorsal to the intestine; they may be attached to the swim bladder (if present). The testes are enlarged before spawning, and white. A **vas deferens** runs along a longitudinal fold in each testis. The two vasa deferentia join in the midline and extend to the **genital pore** just posterior to the anus.

    c. Excise a small piece of the gonad, make a squash preparation of it in seawater, and examine it with a compound microscope. Can you tell the sex of your specimen?

12. The **kidneys** are paired masses that lie against the dorsal body wall and extend the entire length of the abdomen. The caudal ends contain **urinary ducts** that extend from the kidneys to the **urinary bladder.** The bladder lies at the caudal end of the abdominal cavity between the gonad and the swim bladder.

Pericardial Cavity

PROCEDURE

1. Extend the midventral incision to the jaw and expose the **heart** (Fig. 10.5). Enlarge the opening by removing a triangular piece of body wall on each side of the cut. The pericardial cavity is separated from the abdominal cavity by a **transverse septum** (not homologous to the diaphragm in mammals). The heart (Fig. 10.6) has two chambers, a thin-walled **atrium** and a muscular **ventricle.** Blood collected from the venous system enters the **sinus venosus,** a thin-walled

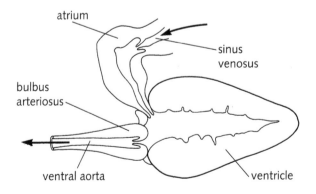

Figure 10.6 Diagram of midline section through heart of bony fish. Arrows show the direction of blood flow.

sac adjoining the atrium posteriorly. Blood then flows into the atrium, and from the atrium to the ventricle. The ventricle pumps blood into a short, swollen **bulbous arteriosus**—the first part of the ventral aorta. Blood is oxygenated as it passes through the gills (Fig. 10.5). It then enters the systemic circulation via the dorsal aorta.

2. If you are working with a fresh specimen, you may be able to prepare a blood smear. Place a drop or two of blood (from the heart or gill) on a clean slide, add a cover slip, and examine it with a compound microscope. Are the blood cells nucleated? Based on the color of the blood, would you say that the oxygen transport pigment is hemoglobin?

Interpretation

1. What is myoglobin's function in red muscle? Briefly indicate how it relates to locomotion in fishes.

2. Name two functions of the swim bladder. What happens if a fish (with a swim bladder) ascends rapidly from a depth of 200 m to the surface (as occurs in commercial fishing)?

3. Based on your observations of the anatomy of your specimen, draw conclusions as to

   a. Its ability to swim and maneuver;

   b. Whether it is pelagic or demersal;

   c. Diet; and

   d. Metabolism (rapid versus sluggish). Defend your answer.

Fouling Communities

OBJECTIVES

After completing this unit, you will be able to

- Recognize the ecological interrelationships of species in a fouling community;

- Observe the morphological adaptations of organisms in such a community;

- Make a quantitative assessment of a community by objectively gathering, recording, and analyzing numerical data; and

- Comprehend how quantitative data can be used to describe biological and ecological aspects of the community under investigation.

INTRODUCTION

Sessile benthic animals and plants must successfully attach to, or penetrate, a substratum while maintaining positions that permit feeding and gas exchange or photosynthesis. The nature of the substratum is the fundamental selective force that determines the life habits and morphology of benthic organisms.

Attached organisms appear on nearly all submerged objects or structures in the sea—ship bottoms, underwater pilings, pipes, bulkheads, ropes, crab traps, shells, and pebbles (Fig. 11.1). This growth is called a "fouling community," in reference to organisms that are unwanted (by humans) because they interfere with efficient functioning of the structure. For example, a ship is designed with a smooth hull to reduce frictional drag as the ship moves through the water. The growth of barnacles on the hull increases drag, thereby impeding its movement through the water. Fouling communities have long concerned humans, who have spent a considerable amount of money and time removing them, and who continue to invest large sums of money and effort to prevent the settlement of such organisms on surfaces that come in contact with the

Figure 11.1 Barnacles attached to the underside of a wooden structure from a harbor on Cape Cod, Massachusetts. Note also the presence of flat, white disks attached to the surface of the wood. The disks are calcareous basal plates, the remains of barnacles formerly growing at these locations.

sea. The freshwater environment also supports fouling organisms, the most notorious (in recent times) of which is the Zebra mussel.

Marine fouling communities are highly diverse; besides algae, more than 1000 species of **macroinvertebrates** have been identified in them (Table 11.1). The most frequently occurring phyla are: Porifera (sponges), Cnidaria (hydroids), Annelida (polychaete worms), Arthropoda (barnacles, amphipods, crabs), Mollusca (mussels, clams, oysters, limpets), Bryozoa (= Ectoprocta, encrusting moss animals), Echinodermata (seastars, serpent stars), and Chordata (tunicates).

Although some fouling species exhibit varying degrees of mobility, the predominant forms live permanently attached (sessile) to a substrate. These sessile organisms exhibit a variety of attachment methods. Solitary epibenthic organisms are often permanently attached to the bottom by **holdfasts** (stalked barnacles, crinoids, seaweeds), **roots** (sea grasses), and **cemented** structures (oysters, acorn barnacles, serpulid polychaetes). Bushlike and vine-shaped colonial forms (sponges, ectoprocts, hydrozoans) also permanently attach to a surface with holdfasts. Sheetlike colonial forms (encrusting ectoprocts, colonial tunicates) are cemented to the surface. Such organisms are not confined to the "fouling" communities of boats and pilings; they may even cover the fronds of submerged aquatic plants (see Unit 6).

Because fouling communities experience large temporal changes in species composition and abundance, studies on fouling communities are often performed for several months, if not years. Seasonal fluctuations indicate temporal mobility in reproductive cycles, settlement, and recruitment. Temporal variability in reproduction is related to seasonal changes in light intensity, nutrients, and availability of food. Epifaunal communities on both the Atlantic and Pacific coasts undergo seasonal changes.

Fouling communities are classic models for demonstrating succession in which: (1) a community goes through an orderly sequence of changes, and (2) a given community is replaced by successive communities over time, ultimately leading to a climax community. Studies of the development of attached communities have documented that microorganisms are the first to settle on new or denuded surfaces. It has been suggested that the formation of a slime film of diatoms or bacteria (or possibly particulate organic matter) is a necessary early step in the process. The slime film apparently renders the surface receptive to the settlement and growth of larval barnacles, mussels, bryozoans, hydroids, sponges, and algae. The microorganisms change the surface–water interface, provide food for some benthic larvae, and possibly afford some protection to the community from toxic elements in the paints and coverings applied by humans to protect the surfaces.

Availability of larvae and their settlement influence the temporal and spatial complexity of the adult fouling community. Most invertebrates of fouling communities have free-swimming larval stages (meroplanktonic) that can be carried long distances by currents. Although this characteristic promotes dispersal of the species, larvae can select particular substrates for settlement and metamorphosis. Settlement often occurs near adults of the population, resulting in gregarious accumulations, as exhibited by barnacles, corals, oysters, and tubeworms.

Because of gregarious tendencies, the density (number of individuals per unit area) of fouling communities quickly increases. This growth reduces the availability of bare substrate, and prevents or retards settlement by new larvae. Competitive interactions for space eventually determine the abundance and distribution of organisms in the community.

The principal strategy used by fouling species to overcome space limitation is overgrowth, achieved by rapid growth exceeding that of neighboring species. Other strategies include the use of chemical secretions to immobilize or kill adjacent species and to prevent settlement by larvae of other species. An oyster bar of the Chesapeake Bay, with its thriving community of epifauna, represents the ultimate example of a sessile community of organisms using one another as a place of attachment.

Boring into hard substrates takes place by mechanical abrasion and chemical weakening of the substrate. Boring sponges, such as *Cliona* sp., maintain a sheet of living tissue on the surface while boring into calcareous substrates.

Boring is used for predation by several types of animals: drilling turbellarians; snails; rock boring bivalve mollusks, such as piddocks (*Pholas* spp.) and shipworms (*Teredo* spp.); and gribbles (isopod crustacean, *Limnoria* spp). Shipworms use the two shell valves as abrasive tools, rotating the shells against the head of the tunnel and rasping away bits of wood. Shipworms construct long, narrow burrows deep in the wood; gribbles make superficial tunnels. Gribble damage is conspicuous; shipworm damage is less visible, but usually more devastating. Some sea urchins, chitons, and limpets are also known to excavate holes in limestone rocks.

Fouling communities are not restricted to sessile epifauna. Growths of sessile forms—such as hydroids, sponges, tunicates, bryozoans, barnacles, mussels, oysters, and tube-building amphipods

Table 11.1 Checklist of commonly occurring organisms in coastal fouling communities. Common genera from Atlantic, Pacific, and Gulf coasts.

Phylum	Class/Order	Common Name	Genus
Chlorophyta		Sea lettuce	*Ulva*
		Tufted seaweed	*Cladophora*
		Hollow seaweed	*Enteromorpha*
Rhodophyta		Tapered red weed	*Agardhiella*
		Coarse red weed	*Gracilaria*
Phaeophyta		Brown fuzz	*Ectocarpus*
Porifera		Boring sponge	*Cliona*
		Bread sponge	*Halichondria*
		Redbeard sponge	*Microciona*
Cnidaria	Hydrozoa	Feather hydroid	*Hydractinia*
		Fern hydroid	*Sertularia*
		Pink hydroid	*Tubularia*
	Anthozoa	Pink anemone	*Aiptasiomorpha*
		White anemone	*Diadumene*
		Brown anemone	*Metridium*
			Aiptasia
		Sea whip	*Leptogorgia*
Platyhelminthes	Turbellaria	Oyster flatworm	*Stylochus*
Rhynchocoela (Nemertea)		Ribbon worm	*Micrura*
Nematoda		Round worms	
Bryozoa (Ectoprocta)		Dead man's fingers	*Alcyonidium*
		Bushy bryozoan	*Bugula*
		White crust	*Membranipora*
Annelida	Polychaeta	Scale worm	*Harmathoe*
		Clamworm	*Nereis*
		Fan worm	*Hydroides*
			Hypsicomus
		Feather duster	*Sabella*
		Sandbuilder	*Sabellaria*
		Mudworm	*Polydora*
Mollusca	Bivalvia	Oyster	*Crassostrea*
		Hooked mussel	*Brachidontes*
		Edible blue mussel	*Mytilus*
		Shipworm	*Bankia*
		Piddock	*Diplothyra*
	Gastropoda	Slipper shell	*Crepidula*
		Periwinkle	*Littorina*
		Drill	*Urosalpinx*
			Thais
		Sea slug	*Doris*
	Polyplacophora	Chiton	*Chaetopleura*
Arthropoda Pycnogonida		Sea spider	*Tanystylum*
			Anoplodactylus

Table 11.1 (Continued)

Phylum	Class/Order	Common Name	Genus
Arthropoda Crustacea	Cirripedia	Ivory barnacle	*Balanus*
		Star barnacle	*Chthamalus*
		Goose barnacle	*Lepas*
	Isopoda	Sea roach	*Ligia*
		Pill bug	*Sphaeroma*
		Common isopod	*Idotea*
		Wood-boring isopod	*Limnoria*
Arthropoda	Amphipoda	Skeleton shrimp	*Caprella*
		Scud	*Gammarus*
		Tube-builder	*Corophium*
	Decapoda	Mud crab	*Rhithropanopeus*
		Glass shrimp	*Hippolyte*
Echinodermata	Asteroidea	Common star	*Asterias*
	Ophiuroidea	Serpent star	*Ophiothrix*
Chordata		Star tunicate	*Botryllis*
		Sea grape	*Molgula*
		Rough sea squirt	*Styela*
		Orange tunicate	*Ecteinascidia*

and worms—provide refuge for mobile species—caprellid amphipods, crabs, shrimp, nudibranchs, polychaetes, and flatworms.

This unit exposes you to minihabitats of the marine environment. Many times we focus our attention on the larger, more conspicuous marine organisms so that we miss the microcosms of activity that exist in the less conspicuous, and sometimes cryptic, habitats. You may have access to one or more of the following communities: oyster community; oyster and clam shells; collections stripped from wharf pilings; "string cultured" communities from coastal waters; or limestone (coral rock).

You will have the opportunity to make a qualitative assessment of the community—that is, what organisms are present, and in what association with one another. That study can be expanded to a semi-quantitative assessment, in which you determine relative abundance of the species. A full quantitative assessment involves the calculation of indices of species diversity and relative importance (see Appendix II).

A. Qualitative Assessment of a Fouling Community

EXERCISE 1

Determining Composition of a Fouling Community

PROCEDURE

1. Work in teams according to directions from your instructor.

2. Obtain a sample of the fouling assemblage in a clean container filled with seawater. Do not disturb the sample until you return it to your study area.

3. The sample may be the entire contents, or a subsample of, a field collection. Treat your sample as though it represents a unique sample, separate from those being examined by others in the class. Record the sample (station) number assigned by the instructor, and any pertinent collection data (for example, date, location, water temperature, salinity, dissolved oxygen) in your notebook.

4. Sketch the undisturbed assemblage, and describe shapes, textures, and relative sizes of conspicuous organisms.

   a. Label the sketch as completely as you can.

b. Provide a scale bar to indicate actual size of the material.

c. Add details to your initial drawing as you work.

d. Place lengthy comments, observations, and additional sketches on separate pages and append them to your report.

5. Use the Key to Macroscopic Coastal Marine Invertebrates (Unit 2) to identify the organisms in your sample. Have your instructor verify your identifications. Arrange your identifications in the same order as those listed in the checklist of fouling organisms in Table 11.1.

a. Begin by examining the sessile erect, filamentous, and encrusting forms.

b. Note the presence of organisms growing on one another. Observe the position of each kind of organism. As you examine these assortments, note the associations that exist between and among the various organisms, and the adaptations involved in this way of life. Consider, as you work through the collection, the advantages and disadvantages of the lifestyles that you observe. Note location and direction of attachment—for example, are barnacles more often attached to the hinge or to the shell margin of mussels? Are they all oriented in the same direction? What species cover others? To what extent?

c. Note the existence of motile forms in, on, and between the sessile organisms.
NOTE: If your study is confined to the Qualitative Assessment, answer the required interpretative questions at the end of this exercise. If you plan to do a Quantitative Assessment, continue with the directions in Part B.

Interpretation

1. Did you observe any autotrophs in your community? What might their presence or absence indicate about the depth from which your community was collected?

2. In your judgment, what was the dominant taxon? Explain how you made this assessment. What does the dominance of that taxon reveal about the community? Is your community relatively young or old in the succession sequence?

3. Were stalked or encrusting forms more prevalent? Could the physical environment (for example, prevailing waves, currents) from where the community was collected have any bearing on this dominance?

4. Describe the principal trophic level of this community (for example, source of food, feeding mechanism).

5. From what depth was your sample collected? Would dominant species differ if the community was collected at shallower or deeper depths?

B. Quantitative Assessment of a Fouling Community

A community can be studied quantitatively to varying degrees. Two approaches are described in the next two exercises. The Semi-Quantitative Assessment involves the identification and determination of relative abundance of the organisms, by numerical count or by surface coverage. Counts and coverage may be estimated. The Full Quantitative Assessment requires greater accuracy in data gathering and analysis, calculation of indices, and graphical plotting.

EXERCISE 1

Gathering Data for Semi-Quantitative and Full Quantitative Assessment

PROCEDURE

1. Plan to record your data and observations in your notebook. Examine the condensed Report Forms A, B, and C.

a. Marine organisms can be anesthetized by placing them in 0.36 M $MgCl_2$ (34.28 g/L of anhydrous $MgCl_2$) for a few hours. This solution is isotonic to seawater with a salinity of 34.6 o/oo. An alternative method is to re-frigerate (5 °C) the sample for a few hours prior to preservation or observation.

b. When the animals are inactive, fix the sample in 10% formalin if it is to be stored for any length of time. If you will study the sample immediately, you may not need to use a preservative. Try not to disturb the order of the community any more than necessary during this procedure.

2. Gather an assortment of containers (small and large petri dishes, finger bowls, Syracuse dishes). Place some seawater in each if you are working with a live sample.

3. Systematically sort through the assemblage and remove the organisms, but do not initially separate any that are attached to each other (for example, sponges lying on mussels).

a. Place the specimens in the small dishes, keeping similar kinds together.
NOTE: Do not destroy the substrate. Certain forms, such as barnacles and encrusting ectoprocts (= bryozoans), cannot be easily removed from the surface of an object. Do not attempt to remove them.

b. Arrange the dishes so that similar taxa are clustered together. For example, if you have four dishes of polychaetes, place them next to one another so that you can list their identifications next to one another on the report forms.

4. Identify the organisms to the lowest taxon possible. Try to distinguish between species, at least designating different species by appearance. For example, two polychaetes might be designated as red nereidlike and black nereid-like polychaetes.

5. Record your identifications in column A of your notebook (see Report Form A). Indicate, in the "comments" column, if the species was mobile or sessile, how attached, and to what.

6. Count the specimens in each dish (each dish has a single species, or taxon) and record this information (see Report Form A, column B, n_i). Numerical abundance is an important component in determining species diversity of the community and for establishing the relative importance of each species to the community.
NOTE: In the case of encrusting forms that cannot be removed from the substrate (see Step 3), count each encrusting mat (the entire colony) as an individual.

7. Drain the fluid from each container, and blot excess moisture from the specimens. Weigh the collective group of specimens, and record the weight in column C (m_i) adjacent to the identified taxon. Be sure to indicate the unit (g, mg) of weight that you are using. This value reflects the biomass of each taxon (species); it is important in determining the relative importance of the species to the community.
NOTE: Certain specimens might be so small and so few in number as to prevent a measurable weight from being recorded. These organisms should be assigned a weight equivalent to the sensitivity of the balance to allow for calculation of index values in the report. For example, if the balance permits you to measure to the nearest 0.1 g, use that as the maximum weight for the "trace" weight. Using the notation "tr" with less than 0.1 g acknowledges that such organisms are present, but in very small quantities.
NOTE: In the case of encrusting forms that resist removal from the substrate (see Step 6), try to estimate the weight (m_i, column C) either by subsampling (if any part of the colony can be removed) or by arbitrarily assigning a weight to it. The assigned weight should not exceed the smallest weight recorded for all other specimens.

8. By this time, the surface of the substrate has been cleared of attached organisms (or they have been measured and recorded). Take measurements (length, width, height, depth, diameter) that will allow you to calculate the surface area and/or volume of the substrate (rock, shell).

a. Estimate the **surface area** of the object (A_o). If it is irregular, trace its outline on a piece of metric graph paper. The area (in cm^2 or mm^2) can be determined by counting the squares within the outline. Remember—a shell has an interior surface as well as an exterior one. Record the information on Form D (Quantitative Assessment: Summary).

EXERCISE 2

Calculation and Data Analysis: Full Quantitative Assessment

PROCEDURE

1. Complete Report Form A, or its equivalent in your notebook.

   a. Calculate the numerical density in column D (number per unit area of substrate = n_i/A_o), and biomass density in column E (weight per unit area of substrate = m_i/A_o).

   b. Total the specimen count for your station ($N_S = \Sigma\, n_i$), and biomass ($M_S = \Sigma\, m_i$).

   c. Tally the total number of species (spp) recorded for your station.

2. Transfer the data of your community (your station) (columns A, B, and C from Form A) to the equivalent of Form B (Diversity and Index of Relative Importance), and share your data with the class (via the blackboard). Gather data from all other stations and enter them on Form B.

   a. Complete Form B by performing the necessary additions.

3. Calculate diversity indices (H and D) for your station using the data on Form A. Refer to the sample set of calculations provided in Appendix II.

   a. Record your values of H and D in the spaces provided at the bottom of Form B, and on Form D. Share your values with the class.

   b. Record diversity indices for other stations examined by the class at the bottom of Form B.

4. Calculate the Index of Similarity for all pair combinations of "stations" examined by the class. Treat each sample as a separate station. Sample calculations are provided in Appendix II.

   a. Record the similarity data in the table provided on Form D.

5. Calculate the Index of Relative Importance (IRI) for each of the taxa listed on Form C. The IRI is based on three parameters: Numerical Composition (NC = percentage of total number of organisms in the sample), Gravimetric Composition (GC = percentage of total weight of all organisms in the sample), and Frequency of Occurrence (FO = percentage of stations inhabited by the species in question). A sample computation of IRI is described in Appendix II.

   a. Compute the NC of the total population for each taxon (column K); NC = $(N_i/N_t) \times 100$.

   b. Compute the GC of the total population biomass for each taxon (column L); GC = $(M_i/M_t) \times 100$.

   c. Compute the FO of the total population for each taxon (column M); FO = $(\Sigma\, s_o/S_t) \times 100$.

   d. Calculate IRI (column N); IRI = (NC + GC) $\times$ FO.

   e. Rank the taxa from highest to lowest according to their IRI values (column O).

Interpretation

Submit your completed report forms, along with responses to the Interpretation questions from Exercise 1, and responses to the following questions (as specified by your instructor).

1. Present your data in visual format (consult Appendix II for guidelines in designing graphs). Include a title (figure legend) for each illustration and provide at least a paragraph for each one, explaining the trend, pattern, and other characteristics.

   a. Rank the species/taxa, as observed by the entire class, on the basis of the Index of Relative Importance (IRI). The graphs described in step 1, b–d, below are based upon the six highest ranked (by IRI values) species/taxa.

   b. Construct **bar graphs** that illustrate (1) percentage numerical abundance (NC), (2) percentage biomass (GC), and (3) frequency of occurrence (F.O). Study the data carefully prior to making the graph. Decide whether you should rank the groups in descending order by count or biomass or arrange them phylogenetically.

   c. Plot the IRI values of the six highest ranked species. The x axis should contain the FO values. The highest ranked taxon will be represented first on the x axis (refer to Report Form C); the lowest ranked taxon will be the last entry on this axis. The y axis represents the GC and NC values. Refer to the example in Appendix II.
   NOTE: If you have a wide range of NC and GC values among your data, it may be difficult to fit the data on the axis of linear graph paper. You might try using a logarithmic scale on the ordinate of a 2- or 3-scale semi-logarithmic graph paper.

d. Construct **bar graphs** for the following data from Form A:

    1. Number/unit area of substrate (column D).

    2. Weight/unit area of substrate (column E) of material. The variable used in your data collection represents the ordinate (*y* axis); the species should be arrayed (according to IRI ranking) along the abscissa (*x* axis).

    3. Include a title for each graph and describe the trend or pattern exhibited by the figures.

    4. How does ranking by substrate surface coverage compare with that based on IRI ranking? Explain the difference.

2. How does your station compare with other stations sampled by the class? Use diversity indices, and the Similarity index to support any conclusions you make.

3. Discuss the merits of using different parameters to describe the relative importance of each species to the community. Compare numerical abundance, biomass, proportionate values (such as weight/unit area), and IRI. Consider the following situations.

a. One species is represented by a single, large organism (for example, one large oyster).

b. An organism's weight is largely in the form of external exoskeleton (for example, barnacles).

c. A species is numerically abundant (hundreds of individuals), but the total biomass is extremely small (for example, caprellid amphipods).

d. Using the literature resources available to you, determine the feeding type (for example, suspension, deposit, predator) of at least the six dominant species in the collection. What is the prevalent feeding type of the community? Can you draw any conclusions from such information as to the specific location or habitat of the fouling community? Explain.

REPORT FORM A

QUANTITATIVE ASSESSMENT

See Table 11.2 for definition of symbols used on this form.

(A) Species	(B) n_i	(C) m_i	(D) n_i/A_o	(E) m_i/A_o	Comments
1					
2					
3					
4					
5					
6					
7					
8					
9					
10					
11					
12					
13					
14					
15					
16					
17					
18					
19					
20					
Total for Station =	N_S _____	M_S _____			

No. Species (spp) =

NAME _____ DATE _____ NO. STATIONS (S_t) = _____

REPORT FORM B

DIVERSITY AND IRI

See Table 11.2 for definition of symbols used on this form.

Species	Stations												Species Total	
	1		2		3		4		5		6			
	n_i	m_i	n_i	m_i	n_i	m_i	n_i	m_i	n_i	m_i	n_i	m_i	N_i	M_i
1														
2														
3														
4														
5														
6														
7														
8														
9														
10														
11														
12														
13														
14														
15														
16														
17														
18														
19														
20														
Totals for Station =	N_s	M_s	N_s	M_s	N_s	M_s	N_s	M_s	N_s	M_s	N_s	M_s	N_t	M_t
(Spp) =														
H =														
D =														

___ ___ + ___ ___ + ___ ___ + ___ ___ + ___ ___ + ___ ___ = ___ ___

REPORT FORM C

IRI

See Table 11.2 for definition of symbols used on this form.

(A)	Total (B)	(C)		NC (K)	GC (L)	FO (M)	IRI (N)	Rank (O)
Species	N_i	M_i	s_o	% N_t	% M_t	% S_t		
1								
2								
3								
4								
5								
6								
7								
8								
9								
10								
11								
12								
13								
14								
15								
16								
17								
18								
19								
20								
Total	N_t	M_t						

No. Species (spp) =

Table 11.2 Definition of symbols used in this unit.

A_o = surface area of substrate (object)

A_S = total surface area of coverage of all species at each station; $A_S = \Sigma\, a_i$

D = diversity index; $= 1 - \Sigma\, (n_i/N_S)^2$

FO = frequency of occurrence, $= (\Sigma\, s_o/S_t) \times 100$; (see IRI)

GC = gravimetric composition, $= (M_i/M_t) \times 100$; (see IRI)

H = diversity index; $= -\Sigma\, (n_i/N_S)\, \ln\, (n_i/N_S)$

IRI = index of relative importance, $= (NC + GC) \times FO$

ln = natural logarithm

m_i = weight of all individuals, each species at each station

M_i = total weight of each species at all stations

M_S = total weight all specimens (all species) at each station; $M_S = \Sigma\, m_i$

M_t = total weight of individuals of all species at all stations; $M_t = \Sigma\, M_S$

NC = numerical composition, $= (N_i/N_t) \times 100$; (see IRI)

n_i = number of individuals of each species at each station

N_i = total number individuals of each species at all stations

N_S = total number individuals all species at each station; $N_S = \Sigma\, n_i$

N_t = total number of individuals of all species at all stations, thus $N_t = \Sigma\, N_S$

SI = similarity index; $= 2C/(A + B)$

s_o = number of stations where each species was found

S_t = total number of stations sampled

Spp = number of species at any given station

REPORT FORM D

QUANTITATIVE ASSESSMENT: SUMMARY

Area Measurements for   Rock   Shell   Wood   Other
Surface Area (A_o) _____ cm^2

Diversity Indices: Calculations based upon your community; data from Report Form A. See Appendix II for explanation and sample calculations.

Shannon–Wiener $(H) = - \Sigma (n_i/N_S) \ln (n_i/N_S)$
$$H =$$

Simpson's $(D) = 1 - \Sigma (n_i/N_S)^2$
$$D =$$

n_i = count of individuals in each species; N_S = sum of n_i for each station (sample); ln = natural logarithm.

Similarity Index: $SI = 2C/(A + B)$

C = number species in common on both stations A and B
A, B = number species on stations A and B, respectively

Table of SI for Six Stations. Data from Report Form B. Expand table as needed. See Appendix II for further explanation.

Station	A	B	C	D	E	F
A	—					
B	—	—				
C	—	—	—			
D	—	—	—	—		
E	—	—	—	—	—	
F	—	—	—	—	—	—

UNIT 12

Sampling the Seashore

OBJECTIVES

After completing this unit, you will be able to

- Understand basic methods used in ecological studies of the seashore;

- Observe the distribution and zonation of organisms inhabiting the seashore; and

- Appreciate the relationship between environmental factors and the observed zonation.

INTRODUCTION

Rachel Carson, in *The Edge of the Sea*, writes:

> The edge of the sea is a strange and beautiful place. All through the long history of Earth it has been an area of unrest where waves have broken heavily against the land, where the tides have pressed forward over the continents, receded, and then returned. For no two successive days is the shore line precisely the same. Not only do the tides advance and retreat in their eternal rhythms, but the level of the sea itself is never at rest. It rises or falls as the glaciers melt or grow, as the floor of the deep ocean basins shifts under its increasing load of sediments, or as the earth's crust along the continental margins warps up or down in adjustment to strain and tension. Today a little

more land may belong to the sea, tomorrow a little less. Always the edge of the sea remains an elusive and indefinable boundary.

In North America, six coastal habitats are accessible by a shore-based sampling approach: rocky coast, sandy beach, salt marsh, tidal flat, dockside fouling community, and sea grass bed. In some coastal areas, these habitats are found within walking distance of each other. In other regions, a rocky habitat may not exist or may be limited to artificial jetties or breakwaters. In certain tropical locations, coral reefs may exist only a short swim from shore. Because it is unlikely that you will be exposed to all of these habitats in one course, plan to take full advantage of the sites that you do visit. The general principles associated with one coastal habitat are applicable to the others; only the specific details associated with differences in terrain vary.

During your field studies you will become familiar with the habitats by surveying the terrain, assessing environmental parameters, and describing the biological communities. Suggested field projects are described briefly in Part B of this unit. Ecological procedures and steps fundamental to all shore studies are described in Part C. Some procedures may be modified or presented in more detail for certain habitats. Details relevant to your specific study sites and projects will be provided by your instructor.

A. Pretrip Planning and Preparation

We must not destroy habitats as we study them. Therefore, during our field studies we will attempt to:

- Avoid destructive sampling whenever possible;

- Identify organisms and determine their abundance in the field;

- Avoid removal of large numbers of specimens;

- Restore anything temporarily disturbed, such as stones and seaweed clumps, to their original sites; and

- Avoid heavy repeated sampling of specific areas.

The Team Approach

Pretrip planning is critical to field work. The success of the trip depends on all participants being well prepared and informed. Read as much as you can about the target habitat, and know the trip objectives.

Be ready to perform effectively and *safely*. Review the section on safety in the General Introduction of this manual, and become familiar with the planned tasks. Pay attention to detail and to all instructions. In your work as part of a team, you may be expected to perform one of the following routine tasks:

- **Recording Angel.** Carry pencils, a checklist, collection logs, data sheets, and a notebook. Record instructions, procedures, observations, data, and comments in your field journal or on appropriate collection logs and data sheets. Write labels for specimen bottles carried by the *Ark*.

- **Ark.** Prior to the trip, make certain that all specimen bottles are available and clean. At the field site, carry the bottles, and place each collected organism in a bottle of appropriate size. Verify that the *Recording Angel* has recorded the necessary information about the collection. If living organisms are to be taken to the laboratory for later study, replace the water in the bottles prior to leaving the field site. Upon return to the laboratory, transfer the specimens to designated holding tanks. If necessary, ensure that preservation of the organisms takes place safely.

- **Buckets.** Carry the buckets, and give them, filled with seawater, to collectors (for example, *Quadrat, Sieve, Nets*).

- **Chemistry.** Carry the water chemistry kit. Collect and treat water samples, and measure other parameters (for example, temperature, pH), as needed.

- **Collecting Nets.** Use as directed by the instructor. Place specimens in buckets with seawater. Clean the nets prior to leaving the field site.

- **Quadrat.** Work with *Transect* to locate biological and sediment sampling stations, and collect samples (with *Sediment*) as required.

- **Panhandler.** Carry white pans, forceps, hand lens, and small dip nets for examining collected specimens prior to their release or transfer to the *Ark*.

- **Screen.** Carry the biological sampling screens and assist in separating organisms from sediment samples. Clean the screen prior to leaving the field site.

- **Sediment.** Work with *Quadrat* and *Shovel* to collect, bottle, label, and analyze sediment samples.

- **Seine.** Carry the beach seine and collect samples. Clean the seine prior to departure from the field site.

- **Shovel.** Carry the shovel and trowel and use them (with *Quadrat, Screen,* and *Sediment*) as directed by the instructor.

- **Transect.** Carry the equipment necessary for locating transects and measuring profiles of the habitat. Work with *Quadrat,* and ensure that all measurements and observations are recorded by the *Recording Angel*.

Materials and Equipment

Dress comfortably and warmly; you should also dress to stay dry. In cold weather, wear clothing in layers to provide insulation and to allow you to remove layers in the event that you become overheated. In warmer months, you will need some type of windbreaker (especially if you will be on a boat). Long pants are always desirable, even if they are worn over a bathing suit or shorts. Bring rain gear (poncho, foul-weather suit) and boots. Field trips are seldom cancelled because of inclement weather.

Always wear protective clothing if you are allergic to poison ivy or prone to sunburn. Poison ivy grows in many shore areas; overexposure to sun is always a hazard near water. Follow the advice found on signs at Australian beaches: "Slip on a shirt, slap on a hat, slop on sunscreen lotion."

Use the checklists of personal clothing and equipment when planning your trip (Table 12.1).

A checklist of scientific materials and equipment commonly used on field trips is provided in Table 12.2. Your instructor will tell you which items will be required for each of your trips. Make sure you know how to use each item.

Geomorphology

Coastal embayments and shores are geologically temporary features. Their physical forms result from the interaction of marine and terrestrial processes. Waves, wind, and tidal currents contribute to erosion, transportation, and deposition of sediments. Topography, elevation, and depth of coastal features serve as indicators of the physical forces that have shaped them. These features can be observed on aerial photographs, U.S. Coast and Geodetic Survey charts, and topographic maps of the study area. Note the location of the study site and the topography of surrounding terrain. Enlarged working copies of a chart of the study site may be used for planning and for recording observations; they should be included in trip reports.

Tidal Considerations

On the seashore, you can witness a change in conditions between the fully terrestrial environment above sea level, and the fully marine habitat below sea level (Fig. 12.1). This gradient results from the submergence and emergence of the shore caused by the periodic **ebb** and **flood** of the **tides.**

Tides are a form of wave produced by the attraction of the sun and moon. Three types of tides occur along the coasts of North America. The simplest **(diurnal)**—only one high and one low tide each day—occurs in parts of the Gulf of Mexico. **Semi-**

Table 12.1   Checklist of personal items for field trips.

Clothing	Personal Equipment	Other
_____ Bathing suits (2)	_____ Camera	
_____ Boots or waders	_____ Clipboard	
_____ Foul-weather gear	_____ Dive mask and snorkel	
_____ Gloves (cotton, gardening)	_____ Film	
_____ Hat (broad-brim)	_____ First-aid kit	
_____ Jeans	_____ Flashlight/batteries	
_____ Sneakers	_____ Knife (pocket)	
_____ Socks (extra)	_____ Notebook	
_____ T-shirts (extra)	_____ Pencils	
_____ Windbreaker	_____ Sea-sickness medication	
	_____ Sunglasses	
	_____ Sunscreen lotion	
	_____ Swim fins	
	_____ Water (drinking)	
	_____ Whistle	

Table 12.2   Checklist of field trip materials and equipment.

_____ Bottles, plastic (for biological specimens)	_____ Plumb line
_____ Bottles, water chemistry (BOD or equivalent)	_____ Push-net
_____ Buckets	_____ Quadrat sampler
_____ Charts (nautical, topographic)	_____ Refractometer/salinometer
_____ Checklist of species (prepared by instructor)	_____ Sediment sieves
_____ Clipboard	_____ See-bucket
_____ Dip nets	_____ Seine
_____ Dissolved oxygen meter	_____ Shovel, straight-edge
_____ Distilled water	_____ Sieves
_____ Forceps, blunt, long	_____ Sorting screens
_____ Formalin (preservative)	_____ Spatula
_____ Hand lens	_____ Species list (checklist)
_____ Labels (waterproof)	_____ Stadia rod
_____ Line, nylon	_____ Station markers (stakes)
_____ Line spirit level	_____ Thermometer, encased
_____ Measuring board	_____ Trays, white plastic
_____ Measuring tape	_____ Transect line
_____ Meter stick	_____ Trowels
_____ Pencils	_____ Vials, plastic (for biological specimens)
_____ pH Meter/pH paper	_____ Water chemistry kit (miscellaneous analyses)
_____ Piling scrape	_____ Water sampling bottle (for shallow water)
_____ Plastic bags (zip-lock)	

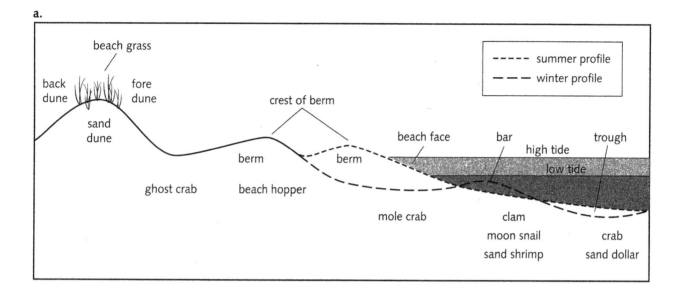

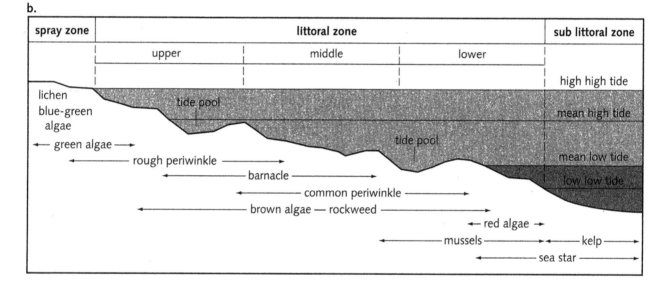

Figure 12.1   Transect profiles of (a) generalized beach, (b) New England rocky shore, and (c) tidal wetland.

diurnal tides of the Atlantic Coast have two highs and two lows each day. **Mixed** tides of the Pacific Coast exhibit diurnal and semidiurnal components; they have two highs and two lows each day, but successive similar tides attain different levels.

Superimposed on the diurnal periodicity of the tides is the regular biweekly cycle of **spring** and **neap** tides. Spring tides have a large amplitude (tidal range); neap tides are smaller in range. A greater amount of the seashore becomes accessible during the low point of a spring tide.

In field research, you must be concerned with both the vertical (**tidal range**) and horizontal water movements (**tidal currents**) of the tides. Knowing the expected tidal range for any field exercise in marine science is important for many reasons, including practical ones. Field trips to shore communities should be scheduled around low tide to ensure maximum exposure of the study sites for collection and observation. Launching and retrieving a boat from a ramp must be planned to allow access to the ramp without danger to the motor vehicle or the boat. When boating, it is vital to know not only the location of shallow reefs and sand flats, but also the height of tide so that these obstacles may be either avoided or traversed safely. Tidal currents also affect boating, and can influence the distribution and zonation of marine organisms.

It is easy to predict the time, height, and current velocities of tides in almost any coastal location in many parts of the world, especially the coasts of North and South America. Consult a current copy of the U.S. National Oceanic and Atmospheric Administration (NOAA) publications: *Tide Tables* (or the companion *Tidal Current Tables*) of the *(East/West) Coast of North and South America*. These annual publications predict the tides (or currents) for an entire year. Using the instructions provided with the tables, you should calculate the tidal heights and estimate tidal currents for your study site as part of the pretrip preparation. It is informative to predict the situation for a three-day period that brackets the day of your collection.

Anticipating the Biological Community

Numerous environmental factors control the distribution and abundance of marine and estuarine organisms. Field trips provide excellent opportunities to observe this relationship.

Spend part of the pretrip preparation reading about the organisms and habitats that you are likely to encounter. Several field guides and reference books can be consulted beforehand (and later, when you write the report). You also may have access to a checklist of organisms known to inhabit the area.

Examining these items before the visit will make it easier to make field identifications and recordings. Tape the checklist in your field notebook (journal).

Field Notes and Collection Logs

You must make critical observations, measurements, and tentative identifications in the field. This information should be recorded in a field journal or collection log. Use a string-bound notebook with lined paper for your journal. Record (in pencil) your field observations on the left page only, keeping the right page blank. This space can then be used to add supplementary notes and ideas at a later time.

Well-designed collection logs serve as guidelines in the field, and they will make it easier for you to interpret your observations when you write your report. Such logs can be inserted in your journal, as necessary. A set of logs is included in this unit. Your instructor may choose to modify them according to need. Scan the logs prior to the trip to gain an overview of what to expect. Any pretrip information or ideas should be written in your journal.

Field Trip Reports

Guidelines for data analysis and report writing are provided in Appendices II and III, respectively. This section includes items for you to consider prior to the trip as well as during posttrip discussions and write-up.

Collaborative effort requires posttrip review. Instructor-guided discussions within each team, or by the entire class, will result in comprehensive notes that form the basis for the field trip report. A single report for each team may be written by the assigned Angels, or an individual report can be written by each member of the class.

Describe each habitat studied. Refer to an appended map or chart. Include such factors as:

- Nature of substratum;
- Degree of exposure to, or protection from, wave action and currents;
- Air and water temperatures;
- Salinity fluctuations;
- Direct sunlight;
- Silt;
- Pollution;
- Tidal effects; and
- Vegetation.

List the most abundant species in each habitat. Identify the few (three or four) most important species that characterize any natural grouping or definable association. If more than one such association exists, describe them.

Record biological observations on individual species, such as:
- Preferred or usual versus less common micro-habitats for each abundant species;
- Tendencies toward consistent orientation relative to each other, to gravity, to light, to wave action;
- Activity or inactivity at the time observed; and
- reproductive condition.

Describe correlations between morphology and physical conditions imposed by the environment.

Consider the area and the organisms studied, their physical and biotic environment, stresses and adaptations. Suggest pertinent questions about the area and attempt to explain the presence and absence of particular species. Some ideas are mentioned in Part C of this unit. Determine if predators, herbivores, scavengers, or suspension feeders are markedly more abundant than others, and offer an explanation for the pattern. Determine what types of plants and animals are poorly represented.

B. Suggested Field Projects and Exercises

Zonation Patterns on Rocky Shores of Various Exposures

Define three or more locations by exposure gradient, degree of seaweed cover, or various coverage by barnacles or mussels. Establish transects and profiles, and mark the stations. Prepare a checklist of species common to the area. Assess the abundance (in quadrats) of the species on the checklist. Estimate coverage of algal canopy. Estimate cover of lichens, mussels, sponges, and other organisms under the algae. Count mobile animals and identify *in situ* wherever possible.

Plot shore profiles, and present zonation/abundance patterns by "kite" diagrams. Abundance may be expressed semi-quantitatively as abundant, common, frequent, occasional, or rare.

Quantitative assessments may be made by presenting results in terms of numbers per m², diversity indices, and indices of relative importance (see Appendix II, and Units 6 and 11).

Infauna Zonation Patterns on Sandy Beaches or Tidal Flats

This study is similar to the rocky shore survey. Select transects and stations on the basis of shore profile, degree of exposure (windward versus leeward beaches on an island, or on either side of a headland), or beach versus tidal flat. Analyze substrate samples for sediments and fauna. Sample water quality. Faunal assessments may be semi-quantitative or fully quantitative.

Floral and Faunal Zonation in Sea Grass Beds

Although similar in approach to the two studies described above, this project requires submerged sampling. Select stations from low tide to as deep as possible. Sample from quadrats randomly placed on an imaginary grid pattern over the sea grass bed.

Determine abundance, relative importance, density, frequency of occurrences, and dispersion patterns of the commonly occurring species, both plant and epifauna. For conservation reasons, you should not attempt infaunal analysis of the grass beds.

Comparison of Fouling Communities

Types and sources of samples can vary. Consult Unit 11 for suggestions and a description of methods for laboratory analysis of the samples. You might compare samples from three different sources, from different heights on a breakwater or piling, or from different substrates. When performing laboratory work, be aware that it may have been difficult to obtain samples that are satisfactory for quantitative assessment. Collecting water quality and tidal data will aid in comparing fouling communities.

Floral and Faunal Zonation in Salt Marshes

This survey involves the use of quadrat sampling, either randomly distributed or along predetermined transects. Select stations from higher elevations to tidal creek beds at low tide. Compare semi-pristine and artificially dredged marshes. Make field identification and counts of conspicuous organisms within the quadrat samples. If appropriate, collect clip quadrats of standing vegetation and infaunal samples at selected sites. Estimate activity of marsh and fiddler crabs by determining the density of burrows along transect lines.

Distribution and Abundance of Shore-Zone Fishes

Select sites that represent different habitats and are amenable to collection with a beach seine, which will easily reveal species composition. Semi-quantitative assessments can be made if the seine is deployed in the same manner (length and depth of tow) at each site. Compare a high-energy beach with a protected shore, beaches representing different salinity regimes along an estuary, or shores with conspicuously different substrates (for example, cobble, sand, mud).

Population Structure Studies

Population structure studies often involve measuring the weight, length, volume, or some other parameter of a group of individuals sampled from a population. The extent of the measurements depends upon the specific objectives of the study and the time available. Such information can reveal the structure of a population in terms of its size–class distribution, or illuminate relationships between the weight and body dimension of a species. Application of these data is described in Appendix II.

1. *Barnacle Population Structure.* Examine the population structure of intertidal barnacles. Determine the frequency distribution of height or basal diameter of the population. Sample and compare different species or populations from different locations. You may expect to see differences related to shore level, wave exposure, and microhabitats.

2. *Shell Form in Gastropod Mollusks.* Depending upon the degree of gene flow, populations may exhibit variation in shell form and color. Drills inhabit the rocky coasts, and then lay eggs in capsules; their young hatch as miniature adults and stay in the same area as the parents. Periwinkles, on the other hand, produce planktonic eggs and larvae, which promote broad gene flow and little genetic isolation.

Sample populations of these species from different locations. Measure the shell width and length of the specimens, and compare the results by location and zone. For example, prepare a scatter plot setting one dimension against the other. Use the differences in the gradient of the lines (fitted by eye) to make comparisons.

3. *Population Structure of Salt Marsh Organisms.* In a manner similar to that described for barnacles, size frequency distributions of bivalve and gastropod mollusks and the presence of crabs collected from different locations within the marsh may reveal differences in zonation by size (age) of the organisms. Differences may also be related to tidal submergence and emergence.

C. Fundamental Ecological Observations

Any stretch of coast separates the marine environment on one side from terrestrial conditions on the other. The coast or shore therefore includes one or more zones that experience both marine and terrestrial conditions to varying degrees. The pattern of the transition influences the arrangement or zonation of organisms that are located in the shore habitat.

The shape of the shore, particularly its cross-sectional profile (Fig. 12.1), results from the combined actions of physical forces, such as wind, rain, waves, and tides. That shape, in turn, influences the extent of the transitional area. These prevailing conditions create the observed biological zonation.

EXERCISE 1

Visual Survey of Terrain

PROCEDURE

Before you enter the study site, examine it visually.

1. Determine the general zonation from conspicuous features, such as plant distribution, strand (wrack) lines, splash zone, water level, and geological formations. Strand lines consist of conspicuous accumulations of seaweed and other debris that wash up on the beach; they identify the upper limits of previous high tides.

2. Look for animal tracks and other signs of recent activity by residents and migrants into the area. Such a survey can identify the presence, abundance, and location of shore birds, water fowl, and gulls.

3. Record your observations in your notebook or on the appropriate collection logs.

4. Choose the locations of **transects** (and corresponding **profiles**). A transect—a section across a shore—is divided into **stations,** or sampling locations, where organisms are studied and counted. A profile reveals vertical features and the slope of the transect.

   a. If the terrain of the study site is generally uniform, as at a broad expansive sand beach, select the transect sites randomly.

   b. If the shore has varied terrain, select several different transects to emphasize differences in zonation. Try to keep each transect on the same type of terrain—for example, all stations on sand versus all on bedrock. If the main purpose is to establish zonation patterns, it is acceptable to shift the transect horizontally.

c. At each site, establish two parallel lines: one for the profile survey, the other for biological sampling. The lines will prevent destruction of biological sampling sites along the transect during profile survey.

EXERCISE 2

Survey of Profiles

PROCEDURE

Using the spirit leveling method (Fig. 12.2), perform the following tasks.

1. Establish a **bench mark** (point zero) at the uppermost location on the shore.

    a. Drive a stake or pipe into the ground.
    NOTE: A rocky coast will require painted marks.
    b. Record the bench mark and all future locations on your map and transect data logs.

2. Estimate and record the position of recent high tides by noting the location of **strand lines.**

3. Run a string (marked off in meters) from the bench mark to the outer boundary of your transect, and tie the end to a stake. Place this stake in the water some distance from the water line, and record its position, the depth of water, the position of the water's edge, and the time. This measurement will reveal height of water at low tide (predicted from tide tables).

4. Measure the vertical elevations and horizontal distances at selected points along the profile (Fig. 12.2a).
   NOTE: The method described below applies to a moderately sloped feature, such as a sandy beach (Fig. 12.2b). Measuring the slope of a bluff or cliff requires modification of the procedure (Fig. 12.2e). First, lay a transect line along the slope and position a meter stick in clear view next to the line to provide a scale. Then, photograph (with slide film) the profile from a known location (bench mark). The resulting slide can be projected onto graph paper and the profile traced (Fig. 12.2d).

    a. Hold one meter stick, or marked pole, at the bench mark and a second one 2 to 3 m downslope (Fig. 12.2a,b). Both poles must be vertical. Attach a spirit level (Fig. 12.2c) to the poles, or use a plumb line, and allow the zero end of the poles to just touch the surface of the substrate.
    NOTE: If the shore is moderately steep or irregular, you may have to place the poles less than 1 m apart to measure differences in vertical height (Fig. 12.2d).

    b. Attach a spirit level to the horizontal pole or transect line (Fig. 12.2a). Lift the pole or line until it is horizontal and stretched between the vertical poles. The line is horizontal when the bubble in the level is centered in the window (Fig. 12.2c).
    1. Measure and record the difference in elevation of the string between the two poles (Fig. 12.2b,d). Subtract the reading on the second pole from that of the first ($= x$ in Fig. 12.2b). A negative value indicates that the second stick is on a decline from the first; a positive value indicates an incline in slope.
    2. Measure and record the horizontal distance between the two vertical poles ($= y$ in Fig. 12.2b).

    c. Move to the next mark and repeat step 4 of this procedure.

    d. Repeat steps 1–4 for other profiles.

5. Compute the cumulative horizontal and vertical distance of each transect line mark relative to the bench mark.

    a. Add (+ elevation changes) or subtract (− elevation changes) all measurements made between the bench mark and the transect line marks.

    b. Record these elevations and horizontal distances on the transect data logs. (An example of correctly computed distances is provided in Table 12.3.)

6. Plot the elevations (on the vertical axis) against the horizontal distances (on the horizontal axis) (see Fig. 12.3). The axes need not have the same scale; an exaggerated scale better reveals slope changes.

7. Mark the tide levels on your graph.

    a. If you made the transect at the time of low tide, show the location of the water level on the graph (consult tide tables for predicted time).

b. If the transect was made at a time other than predicted low tide, estimate the position of low tide on your graph. Determine the position of mean high tide on your graph.

8. Plot other measurements made within the profile/transect on separate graphs using the same vertical scale (see Fig. 12.3).

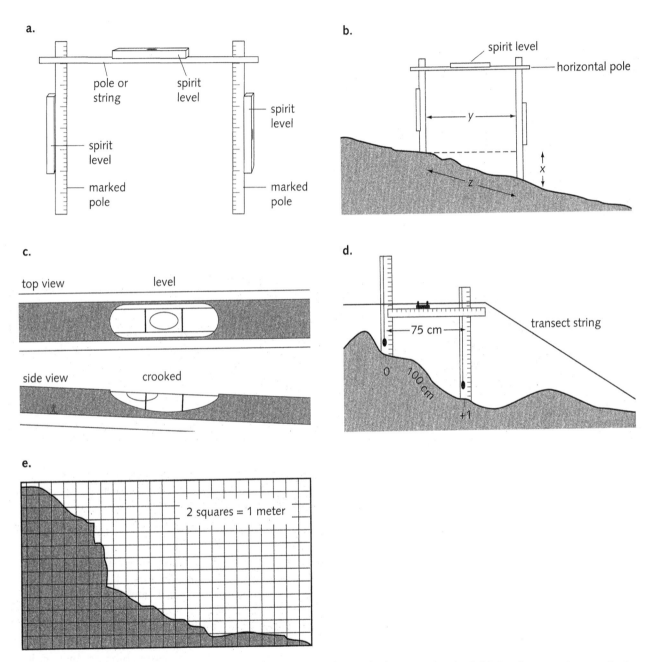

Figure 12.2 Shore profile. (a) Two vertical poles (meter sticks) and a horizontal pole. Spirit levels assure true vertical and horizontal positions. (b) Bench mark and location of two stations are indicated by vertical poles. The elevation change between the two poles is indicated by the difference between height of leveled string on the two poles. For example, readings on the first and second poles level with the horizontal pole are 65 cm and 95 cm, respectively. Difference in height (x) between 1 and 2 = 65 cm − 95 cm = −30 cm, a decline. Horizontal distance (y) is measured directly from the horizontal pole, line, or metric tape. (c) The line spirit level indicates the horizontal position of the measuring string when the bubble is centered in the window. (d) Profiles on irregular terrain can be determined by placing meter sticks close together and measuring distance between them, as well as their vertical difference (= elevation). (e) The profile of a steep slope can be determined by projecting the image from a slide transparency onto graph paper.

Table 12.3 Sample transect data. Data plotted in Fig. 12.3.

Transect Mark No.	Horizontal Distances (m) (y =)	Cumulative Horizontal Distances	Elevation Changes (cm) (x =)	Cumulative Elevation Changes (cm)
0	0	0	0	0
1	3	3	30	30
2	2	5	−40	−10
3	5	10	−20	−30
4	5	15	−20	−50
5	5	20	−30	−80
6	5	25	−30	−110
7	2	27	30	−80
8	3	30	−25	−105
9	5	35	−15	−120

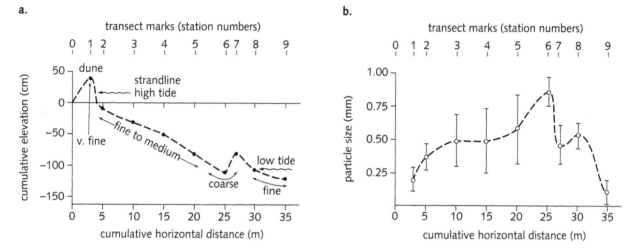

Figure 12.3 (a) A plot of transect data from Table 12.3 provides a visual portrait of the profile. Other notations made on the plot include location of strand line (recent high tide line), level of low tide, general description of substrate types. (b) Range and modal particle size of sediment samples plotted for each station along transect.

EXERCISE 3

Sampling the Biota

It is important to sample the biota as well as corresponding sediments, water samples, and other items. Two kinds of biotic sampling are possible: destructive and nondestructive. In destructive sampling, organisms are removed from the shore for counting, weighing, and identification. Although this provides the most accurate way to quantify the biota, it disrupts the community and should be avoided.

Nondestructive sampling is preferable in most instances. Biomass can be assessed without total destruction. Solitary and/or mobile animals can be

counted by placing a quadrat over a particular area; the count can then be expressed as **density** (no. m^{-2}). Slow-moving animals and sessile plants and animals compete for space. **Percentage cover** of available space can be estimated as an ecologically meaningful measure of abundance. For identification purposes, and for assessing biomass by weight, small samples of organisms can be sampled.

Abundance scales represent semi-quantitative estimates of density or cover that are assigned to several broad categories (Table 12.4). They provide a rapid estimate of a species over a broad area and allow the integration of abundance over an area, which otherwise could only be studied by many quadrats. They are prone to subjective error, however.

Table 12.4 Abundance scales for some rocky shore organisms. Modified from Hawkins and Jones (1992).

Algae

E	>90% cover
S	60–90% cover
A	>30% cover
C	5–30% cover
F	<5% cover (zone still apparent)
O	Scattered individuals (zone indistinct)
R	Few plants—30-min search

Small Barnacles

E	>5 cm^{-2}
S	3–5 cm^{-2}
A	>1 cm^{-2} (rocks well covered)
C	0.1–1 cm^{-2} (up to 1/3 rock covered)
F	100–1000 m^{-2} (individuals never $>$ 10 cm apart)
O	1–100 m^{-2} (few within 10 cm of each other)
R	Few found—30-min search

Large Barnacles

E	>300 per 10 × 10 cm
S	100–300 per 10 × 10 cm
A	10–100 per 10 × 10 cm
C	1–10 per 10 × 10 cm
F	10–100 m^{-2}
O	1–9 m^{-2}
R	Few found—30-min search

Mussels

E	>80% cover
S	50–79% cover
A	>20% cover
C	Large patches
F	Scattered individuals/small patches
O	Scattered individuals/no patches
R	Few seen—30-min search

Limpets

E	>200 m^{-2}
S	100–200 m^{-2}
A	>50 m^{-2}
C	10–50 m^{-2}
F	1–10 m^{-2}
O	<1 m^{-2}
R	Few found—30 min-search

Lichens, Lithothamnia Crusts

E	>80% cover
S	50–79% cover
A	>20% cover
C	1–20% cover (zone well defined)
F	Large scattered patches (zone ill defined)
O	Small, widely scattered patches
R	Few patches seen—30-min search

Dogwhelks, Topshells, Anemones, and Sea Urchins

E	>100 m^{-2}
S	50–90 m^{-2}
A	>10 m^{-2}
C	1–10 m^{-2}, very locally >10 m^{-2}
F	<1 m^{-2}, locally sometimes more
O	Always <1 m^{-2}
R	1 or 2 found—30-min search

Large Periwinkles

E	>200 m^{-2}
S	100–100 m^{-2}
A	>50 m^{-2}
C	10–50 m^{-2}
F	1–10 m^{-2}
O	<1 m^{-2}
R	1 or 2—30-min search

Small Periwinkles

E	>5 cm^{-2}
S	$>$3–5 cm^{-2}
A	>1 cm^{-2} at H.W.N. (extending down to mid-littoral)
C	0.1–1 cm^{-2} (mainly in littoral fringe)
F	<0.1 m^{-2} (mainly in crevices)
O	A few individuals in deep crevices
R	1 or 2 found in 30-min search

Large Tubeworms

A	>500 m^{-2}
C	100–500 m^{-2}
F	10–100 m^{-2}
O	1–9 m^{-2}
R	<1 m^{-2}

Small Tubeworms

A	5 cm^{-2} on >50% of surface
C	5 cm^{-2} on <50% of surface
F	1–5 cm^{-2}
O	<1 cm^{-2}
R	Few found—30-min search

Key:
E = Extremely abundant
S = Superabundant
A = Abundant
C = Common
F = Frequent
O = Occasional
R = Rare
N = Not found (all cases)
H.W.N. = High water neap

PROCEDURE

1. Follow the fundamental rules of sampling.

   a. Study first that part of the shore that will be exposed the shortest period of time. Sampling subtidally is not only possible; it is recommended.

   b. Collection of organisms may be prohibited in specified protected areas. Be prepared to make your identifications and observations of organisms in the field.

   c. If collection is permitted, obtain representative samples of the sediments, plants, and animals from along the transect.
      1. Make tentative identifications of the organisms, and note any special adaptations exhibited at the time of sampling.
      2. Label the samples and place them in suitable containers for later processing.
      3. Record such information relative to station location and number.
      NOTE: Your instructor will advise you which organisms should be preserved, and which are to be kept alive for further study in the laboratory. Place living organisms in buckets containing seawater. Do not place crabs in the same bucket with fish, or crabs and fish in the same one with softbodied invertebrates.

2. Determine how to make observations along the transect.

   a. The simplest method to determine zonation of organisms is to record everything that the transect line touches.

   b. Another method is to lay a meter stick at right angles to the transect line at predetermined intervals and record all the organisms touched.

   c. The quadrat sampling method provides a quantitative assessment of the biomass. It is described below.

3. **Quadrat Sampling.** A quadrat sampler is a square frame of known dimensions (Fig. 12.4). Its size depends upon the type of habitat, the anticipated numerical abundance of the organisms to be sampled, and convenience. For example, it may seem logical to use a quadrat encompassing an area of 1 m². Carrying a large object (1 m on each side) into the field might be awkward, however, and collecting and/or counting every individual within that area would be a formidable task. Instead, the sample size must be reduced to something more manageable—for example, a quadrat that is 0.50 m per side (sample area = 0.25 m²).

   If you find large populations of organisms, it will be necessary to randomly select smaller subsamples within the quadrat. Consider the quadrat to be a two-dimensional grid that is subdivided into squares equal to the area of a sampling unit (Fig. 12.4a,b). The squares on two adjacent sides of the grid are numbered, and a

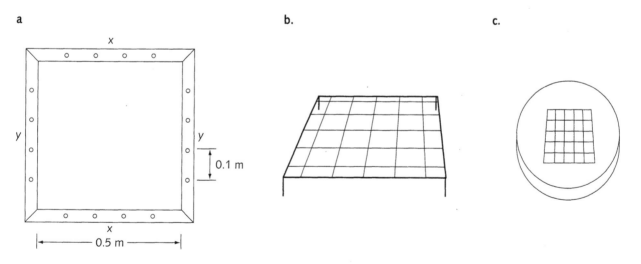

Figure 12.4 Quadrats. (a) Constructed of metrically ruled metal or wood (from meter sticks). (b) With cross-strings functioning as counting points. (c) A 5 × 5 cm quadrat scratched onto lid of a plastic petri dish for subsampling larger quadrats.

pair of coordinates identify each sampling unit. Numbers are drawn in pairs from a table of random numbers (refer to any statistics textbook) or from a pocket calculator with random number generation. These random numbers are used as coordinates to locate each unit in the sample. Remember to perform the correct mathematical transformations during your data analyses as you calculate abundance per unit area.

a. Place a quadrat sampler (Fig. 12.4) at designated intervals along the transect line. Samplers must be located in an unbiased manner (that is, randomly). Do not place them conveniently or selectively on the basis of some inviting location.

b. Determine the sample size. If the individuals are large, increase the sampling area to achieve a meaningful count. If the organisms are very small or very numerous, subsample to avoid impossibly large counts.

c. Study the epifauna and epiflora prior to disturbing the substrate. Record your observations on the epibiota analysis data log.

d. Measure the abundance of **epibiota**, perhaps by using **abundance scale**s (Table 12.4).
    1. Count the number of individuals of each species or taxon in the sampling unit. Remove a representative of each species to verify its identity later. Record the tentative identification in your journal, and label the specimen container accordingly.
    2. Determine the **percentage coverage** of encrusting forms and some plants. Measure the area of the substrate and the area covered by the species.
    3. Determine **biomass.** Remove samples of the organisms within a predetermined quadrat area, and weigh them. Plants are clipped off at ground level, hence the term "clipquadrat." Weight (wet or dry) is used to assess the biomass of the sample. This method is applicable to both plants and fauna. For dry weight, plants are dried to constant weight at 50 °C; higher temperatures (90 °C) are used for animals.

e. Remove **infaunal** and **sediment** samples with a square-edge shovel or hand trowel, depending upon the area and volume of sediment to be sampled (Fig. 12.4b). Record your observations on the infauna analysis data logs.

1. Note any layering by color or texture of the substrate.
2. If quantitative substrate analysis is planned, remove two samples from the quadrat: one for biological sorting (macrofauna and/or meiofauna, see Unit 4), the other for sediment analysis. Place substrate samples in labeled containers for later analysis.
3. If no quantitative substrate analysis is planned, place the material in a sorting sieve and carry it to the water's edge. Wash and sort it immediately. Place the organisms in plastic bags or jars with labels for later analysis in the laboratory.

4. Other Biological Information.

a. Sample unusual or unique permanent objects adjacent to transects. Pilings, for example, often contain an assortment and zonation of fouling organisms (see Unit 11) that are strikingly different from those of the sandy beach in which the pilings stand.

b. Make **seine** collections in the subtidal zone of the study area for qualitative sampling. Because fish can evade the net and the seine provides unequal sampling, this method cannot be used in quantitative assessments.
    1. Two students work the seine. One student remains at the water line while the other wades into the water perpendicular to shore, paying out the net along the way.
    2. When the net is fully extended or the water is waist-deep, both students drag the net parallel to shore. Move the net slowly so that the lead line stays against the bottom and prevents fish from escaping under the net.
    3. After the net has been dragged 20 to 50 m, the student farthest from shore should move faster than the student in shallower water, and begin to drag the net toward shore. Advance the net up the beach, guarding against lifting the lead line off the bottom. Both students drag the seine onto the beach.
    4. Sort the specimens into buckets of seawater, and make tentative identifications. Measure the specimens and make any other observations deemed necessary by the instructor. Record your data on the seine data log. Return all specimens to the water unless you need to retain samples for later identification.

EXERCISE 4
Physical Observations

Record air, water, and substrate temperatures at collecting sites. Note wind direction and velocity, wave height, currents, and their influence on substrates.

EXERCISE 5
Chemical Analyses

Collect and prepare water samples at designated collecting sites. Refer to the exercises on water chemistry (TSDO, Unit 1) for details. Record salinity, dissolved oxygen, and pH.

EXERCISE 6
Sediment Analysis

Particulate materials deposited along the coast and on the bottom of the ocean are called sediment. They influence biotic zonation by controlling the types of organisms that inhabit the benthic realm. Morphology and mode of feeding are intimately related to type of substratum.

Sediment particles range in size from very fine silts and clays, through moderate-size sands and pebbles, to cobbles and boulders (Fig. 12.5a). The relative proportions of these components reveal the nature of the coast, currents, tides, and prevailing winds. Fine particles tend to accumulate in protected coves and embayments featuring little wave action and low current velocity. Larger particles are typically found in much more physically dynamic conditions, such as an exposed sand beach.

Color also offers an important clue to mineral content and the relative amount of dissolved oxygen in the interstitial (between the grains) water of the sediment. For example, muds that are green, gray, yellow, or brown and that do not possess an unpleasant odor indicate the presence of plentiful oxygen. If they are rich in organic matter, the muds are usually black and smell like rotten eggs (hydrogen sulfide). These characteristics point to anaerobic or anoxic conditions (no oxygen in the interstitial water).

Two procedures for sediment analysis follow. The first—a preliminary analysis based on human sensory perception (smell and touch)—can be performed in the field. The second method—a quantitative analysis—must be performed in a laboratory.

Preliminary Sediment Analysis

PROCEDURE

1. Remove a sediment sample and set it aside. Examine the layering of the sediment on the sides of the excavation. Measure the depth of each layer and describe the color of each layer. Use the Preliminary Sediment Analysis Report Form, part **A**.

2. Examine the extracted sample. Remove samples of each layer and describe their odors, part **B**.

3. Rub a bit of each layer between your thumb and forefinger. If it feels gritty, it probably contains sand; if smooth, it is silt or clay; if smooth and sticky, it is largely clay, part **C**.

4. Examine the samples with a hand lens and try to match them to a standard size on the sediment sizing chart (Fig. 12.5a). Estimate the predominant sizes and record your results in part **D**.
   NOTE: Mixtures of substrates can be described as follows: silty clay = mostly clay, some silt, no sand; sandy mud = silt and clay with some sand; silty sand = sand with some silt.

5. Estimate how much of the sample contains shell and plant material, and record your data in part **E**.

6. Be alert for any signs of pollution, such as oil film or unusual odors. Note the presence of any human artifacts in the substrates.

7. If you made a transect of the area and took sediment samples along it, plot information such as dominant sediment type and depth of black layer on a graph that shows the profile of the transect line (Fig. 12.3).

8. If you made several transects, make a map of the study area that illustrates zonal distribution of the principal sediment types. Compare this distribution with the estimates of abundance made during the biotic survey.

Quantitative Analysis of Percentage Sand and Silt–Clay.

PROCEDURE

1. Determine the proportion of dry weight. Use the Sediment Analysis Report Form, lines **A–C**.
   NOTE: To show differences in layers, sample each layer and treat each sample separately.

a.

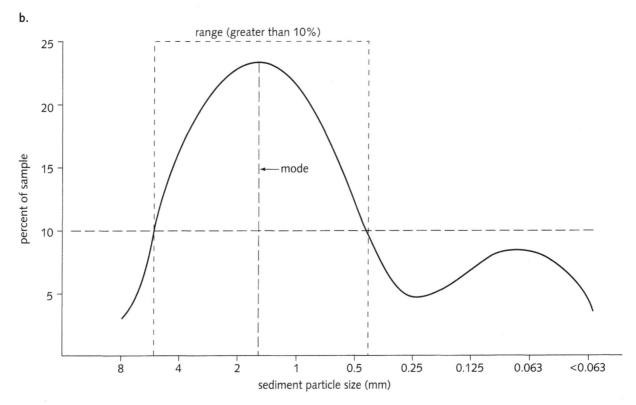

pebbles (64–4 mm) use ruler	cm 1 2 3 4 5 6 7 8 9 10 11 12 13 14 15			
granules (4–2 mm)				
sand	very coarse (2–1 mm)		particle roundness/angularity determined easily by eye	
	coarse (1–0.5 mm)			
	medium (0.5–0.25 mm)			
	fine (0.25–0.125 mm)		each particle easily distinguished by eye	
	very fine (0.125–0.063 mm)		somewhat gritty	
silt (0.063–0.0039 mm)			smooth, not sticky	
clay (below 0.0039 mm)			smooth and sticky	

b.

Figure 12.5 Sediment analysis. (a) Sediment sizing chart for estimating the particle size of substrates. (b) Size frequency curve of sediment particles. Modal particle size of each sample is used to construct the graph shown in Figure 12.3b.

a. Record the weight of a clean, drying crucible or small beaker on line **A**.

b. Weigh exactly 10 g of the sediment sample and place it in the drying container.

c. Dry the sample to constant weight at 105 °C (preferably overnight). Record the dry weight (sample + container) on line **B**.

d. Calculate dry weight proportion. Divide sediment dry weight (line **B** − line **A**) by the sample wet weight (= 10 g). Record the value on line **C**.
   NOTE: This value is a proportion, not a percentage. To obtain a percentage value, you need to multiply line **C** by 100!

2. Determine the sand fraction (lines **D**–**G**).

   a. Oxidize the organic matter. Place exactly 100 g (wet weight) of a sample from the same substrate in a 1-L flask (unstoppered) and add 100 mL of fresh 6% hydrogen peroxide (or 200 mL of 3% strength). Swirl the flask until all lumps are broken up and no more gas forms. Let the flask stand overnight and then **warm** it gently. Add more peroxide until the reaction ceases. Continue to warm the mixture for another 10 min to remove excess peroxide. Stopper the flask and shake it vigorously.
   NOTE: An alternative method involves the use of a muffle furnace. No chemical treatment is involved—instead, the sample is ignited at a very high temperature.

   b. Carefully clean a dry 63-μm sieve with a brush. Record the weight of the sieve on line **D**.

   c. Place the sieve in a basin of water and wet-sieve the sediment sample by gently shaking, rotating, and puddling (lifting up and down) the sieve until most of the fine materials have passed through. Discard the silt–clay remaining in the basin unless you plan to perform other analyses on it.

   d. Dry the sieve and its contents for one hour at 105 °C. Remove the sieve from the oven (with hot pads) and allow it to cool to room temperature.

   e. Hold the sieve over a piece of white paper or a white tray. Vigorously shake it over the paper until no more silt or clay passes through.

f. Record the weight of the sieve and the sand fraction on line **E**.

g. Record the dry weight of the sand fraction on line **F** (= line **E** − line **D**).

h. Determine the dry weight of the sieved sediment sample. Multiply the total dry weight proportion **(C)** by the 100-g wet-weight sample. Record this value on line **G**.

i. Determine the dry weight of the sand fraction remaining on the sieve. Subtract the empty weight of the sieve **(D)** from the total weight of the oven-dried sand fraction plus sieve **(E)**. Record this value on line **F**.
   NOTE: If you will determine the relative proportions of particle sizes within this sand fraction, save the sample and go to the procedure for Particle Size Distribution of Sand.

3. Calculate the dry weight of the silt–clay fraction by subtracting the dry weight of the sand **(F)** from the total dry weight **(G)**. Record this value on line **H**.

4. Calculate the percentage sand by dividing the sand fraction dry weight **(F)** by the total sample dry weight **(G)**, and multiplying the result by 100. Record this value on line **I**.

5. Calculate the percentage silt–clay by dividing the silt–clay fraction dry weight **(H)** by the total sample dry weight **(G)**, and multiplying the result by 100. Record this value on line **J**.

6. Plot your data for percentage sand and percentage silt–clay along the profile of your study area.

Particle Size Distribution of Sand

PROCEDURE

1. Dry about 100 g of sand at 105 °C. Record the exact value.

2. Stack a series of graded sieves together, with the coarsest mesh on top and the finest on the bottom. Place a sieve pan under the bottom sieve. Record the sizes in column **K** on the Sediment Analysis Report Form.

3. Pour the oven-dried sediment sample into the top sieve and place a cover on top.

4. Shake the sieves for 10 to 15 min with an automatic shaker or by rocking and tapping them by hand.

5. Record the mesh size of the first sieve and weigh a clean beaker (column **L**). Empty the sediment from the first sieve into the container. Record the combined weight of sediment and container (column **M**). Subtract the container weight **(L)** from the total weight **(M)** to determine the weight of the sand fraction in that sieve. Record the value in column **N**.

6. Repeat the previous step for each sieve in the series.

7. Determine the percentage of the original sand sample in each sieve. Divide the weight of each fraction **(N)** by the total dry weight of the sediment sample (line **G**), and multiply the result by 100. Record this value.

8. Construct a size frequency curve for each sample. Arrange the sieve sizes along the horizontal axis. The vertical axis is the percentage of sediment in each of the sieve size classes (Fig. 12.5b).

9. Compare different samples by referring to the size of the mode (high point) of the curve generated in the previous step. Plot your data along the profile of your study area. Particle size can be presented on the vertical axis, and your stations along the horizontal axis (Fig. 12.3b).

TRANSECT DATA LOG

Location _____ Date _____ Time _____

Location of bench mark:

_____ meters from landmark _____

_____ meters from landmark _____

_____ meters from landmark _____

Time of mean low water _____

Transect Line Mark No.	
Elevation change + = rise, − = fall (cm) Measured from mark to mark	
Horizontal distance (cm) Measured from mark to mark	
Elevation change + = rise, − = fall Cumulative from bench mark to mark (calculated)	
Horizontal distance (cm) Cumulative from bench mark to mark (calculated)	
Tidal marks ML = mean low MH = mean high	
Sediment particle size mode/range (cm)	
Temp (°C)	
Salinity (o/oo)	
D O (ppm)	
Other measurements	
Other measurements	

NAME _____ SECTION _____ DATE _____

EPIBIOTA DATA LOG

Collection No. _____ Station No. _____

Location _____ Date _____ Time _____

Quadrant Sample/Subsample Area _____

Transect Mark No.	Epibiota Survey (No./m²; % Cover)				
	Species	Species	Species	Species	Species
_____					
_____					
_____					
_____					
_____					
_____					
_____					
_____					
_____					
_____					

Comments: _____

INFAUNA ANALYSIS DATA LOG

Collection No. _____ Station No. _____

Vessel _____ Date _____ Time _____

Depth of Water (m) _____

Bottom Sampler _____ Sample Size (m) _____

Bottom Sediment _____ Sieve Size _____

No. in sample	No. per m^2	Taxon	No. in sample	No. per m^2	Taxon
_____	_____		_____	_____	
_____	_____		_____	_____	
_____	_____		_____	_____	
_____	_____		_____	_____	
_____	_____		_____	_____	
_____	_____		_____	_____	
_____	_____		_____	_____	
_____	_____		_____	_____	
_____	_____		_____	_____	
_____	_____		_____	_____	

NAME _____ SECTION _____ DATE _____

BEACH SEINE COLLECTION LOG

Collection No. _____ Station No. _____

Seine Length _____ Mesh _____ Date _____

Time of Sample _____ Time of Mean Low Water _____

Length of Tow _____

Description of Substrate:

Species	Number/ Weight/Volume	Species	Number/ Weight/Volume
_____	_____	_____	_____
_____	_____	_____	_____
_____	_____	_____	_____
_____	_____	_____	_____
_____	_____	_____	_____
_____	_____	_____	_____
_____	_____	_____	_____
_____	_____	_____	_____
Comments:			

PRELIMINARY SEDIMENT ANALYSIS

Location _____   Date _____

Transect Mark _____   Station _____

A. **Color**
   Record the three most common colors (if sand, record the three most common grain colors)

   1. _____

   2. _____

   3. _____

   Is the color wet or dry?

   Draw a picture of rocks or boulders.

B. **Odor**
   Describe odor in common terms

   _____

   _____

   _____

   _____

C. **Texture**
   Describe texture in common terms

   _____

   _____

   _____

   _____

D. **Sediment Size**
   Record most frequent size (from sediment sizing chart)

   _____

   _____

   _____

   _____

E. **Shell and Plant Material**
   Record average amount of shell and plant material
   Subsamples (shells)

   1. _____

   2. _____

   3. _____

   average _____
   Subsamples (plants)

   1. _____

   2. _____

   3. _____

   average _____

F. **Organisms (Living)**
   Record kinds and numbers of organisms

   1. _____

   2. _____

   3. _____

   4. _____

   5. _____

   6. _____

   7. _____

   8. _____

   9. _____

   10. _____

G. **Pollution**
   Record any evidence of pollution

   _____

   _____

H. **Mineral Identification**
   Record the most common minerals

   1. _____

   2. _____

   3. _____

   4. _____

   5. _____

SEDIMENT ANALYSIS REPORT FORM

Sample Information

Date _____   Location _____   Station No. _____

Transect Mark No. :_____   Sample No. _____   Core No. _____

Core Length _____   Horizon No. _____   Distance from Top of Core _____

Other Information

DETERMINING THE PERCENT SAND AND PERCENT SILT–CLAY OF MUD

A. Weight of empty drying crucible or beaker   _____ g

B. Weight of oven-dried sediment plus drying container   _____ g

C. Dry weight proportion $= \dfrac{\text{dry weight}}{\text{wet weight}} = \dfrac{(B) - (A)}{10 \text{ grams}}$ _____ g

D. Weight of empty sieve   _____ g

E. Weight of sieve plus oven-dried sand fraction   _____ g

F. Dry weight of sand fraction $= (E) - (D)$ _____ g

G. Total dry weight of sediment sample $= (C) \times 100$ grams =
    dry weight of proportion $\times$ wet weight =   _____ g

H. Dry weight of silt–clay fraction $= (G) - (F) =$ _____ g

I. Percent of sand in sample $= \dfrac{(F)}{(G)} \times 100 =$ _____ %

J. Percent of silt–clay in sample $= \dfrac{(H)}{(G)} \times 100 =$ _____ %

SEDIMENT ANALYSIS REPORT FORM (continued)

DETERMINING THE PARTICLE SIZE DISTRIBUTION OF SAND

K.	L.	M.	N.	
Sieve Size (mm)	Empty Beaker Weight	Beaker plus Sand Weight	Sand Weight = (M) − (L)	Percent of Total Original Sample = $\dfrac{(N)}{(G)} \times 100$
				%
				%
				%
				%
				%
				%
				%
				%
				%

UNIT 13

Offshore Sampling

OBJECTIVES

After completing this unit, you will be able to

- Understand shipboard methods of collecting water samples and organisms in coastal waters;

- Observe distribution and zonation of organisms inhabiting the sea bottom and the water column; and

- Appreciate the relationship between water quality and observed biological distribution.

INTRODUCTION

A logical place to perform marine biological research is in or on the water—often far at sea, out of sight of land. Shipboard research is exciting, but it also is hard (and sometimes dangerous) work. Research vessels are unstable platforms that are exposed to saltwater and rough weather, have restricted space usually crammed with apparatus and instrumentation, and are expensive to operate. If you have the opportunity to participate in a research cruise, *carpe diem.*

This unit introduces a variety of basic methods used to collect water samples and to sample the organisms that live in the water column and on (and in) the bottom sediments. The extent to which the instruments and methods are used depends upon the objectives of the research cruise. For example, the cruise may have a strictly "hydrographic" purpose, emphasizing the collection of water quality data (for example, temperature, salinity, dissolved oxygen, transparency). Other cruises may be confined to plankton sampling or trawling.

Pretrip planning, as described in Unit 12, is essential to the success of the operation. Familiarize yourself with the trip objectives, plan of operation, and safety regulations.

Shipboard Safety

Review the safety section in the General Introduction to this manual, and follow these rules:

1. Board or disembark from the boat when you are instructed to do so by the captain (the person who is *always* in charge)! Do whatever the captain requests, *instantly.*
2. Sit, preferably inboard (not on the gunwales— the sides), whenever you are not engaged in sampling. Keep your hands off the gunwales during any docking procedures.
3. Know the location of the life preservers, and how to put them on. You may not be required to wear a life preserver unless conditions warrant, although nonswimmers should wear one at all times.
4. Know the location of life-rings, the first-aid kit, and fire extinguishers.
5. In the event of *man-overboard*, **do not** jump in after the victim. Notify the captain and the instructor immediately. Throw a life-ring to the victim. Keep your eye on the victim until rescue is completed.
6. Report any injuries immediately to the captain and instructor. Use the first-aid kit for small cuts and abrasions. If the injured person is lying on the deck, do not move the patient. Follow the instructions of the captain.
7. Never deploy any equipment from the boat until instructed to do so. Retrieve equipment and store it immediately when so ordered.
8. Be aware of any winches, ropes, cables, and other deck gear used to deploy and retrieve sampling gear. Keep your hands and feet away from any such items in motion. Unless you are explicitly asked to participate in such activities, stay out of the area of activity.
9. Sneakers, deck shoes, or boots are required dress for boating. Bare feet are *forbidden!* Boat

decks are slippery, and bare feet are no match for collecting gear.

10. If you are prone to seasickness, consult a physician to obtain the appropriate medication. Eat a light meal prior to the trip, and keep yourself busy during the journey to and from the sampling sites. Be polite—if you do get seasick, do it on the leeward side of the boat.

Clothing and Personal Equipment

Refer to the checklists of personal clothing and equipment provided in Unit 12.

A. Physical and Chemical Sampling

Initial observations are made as soon as the research vessel (R/V) has reached its destination and sets anchor—that is, when it is "on station." At that time you will be notified of the station number and location. You will perform certain tasks or observe demonstrations that are designed to either measure the water quality directly or retrieve water samples for later analysis in the laboratory. Record your observations in your field journal or on oceanographic data logs (a sample is provided in this unit).

EXERCISE 1

Determining the Transparency of the Water

Sediments, plankton, dissolved pigments, and suspended pollutants decrease the degree to which sunlight can penetrate water. Low transparency readings indicate high turbidity, and indirectly reveal the depths to which phytoplankton can survive and efficiently carry out photosynthesis. Two methods of determining water transparency are described here.

Sampling Equipment and Oceanographic Equipment

Certain instruments used to measure water quality in Units 1 and 12 (for example, pH meters, salinometers/refractometers, and dissolved oxygen meters) are also employed on board research vessels. Some pieces of equipment that are deployed from research vessels to collect water samples, measure physical and chemical parameters, or collect organisms are described below.

PROCEDURE

1. Use a **photometer,** an instrument that directly reads light intensity or percentage transmittance. Because various models are available, you should consult the user's manual for operating instructions.

2. Measure transparency with a **Secchi disk** (Fig. 13.1). A Secchi-disk reading provides the depth of water at which the naked eye can no longer see the disk. Such readings give a visible sense of water transparency, but do not measure light intensity directly. However, you can estimate the degree to which surface light intensity is reduced in the sampled water column. You can estimate the depth of the photic layer of the water sampled (about 2.5 times the Secchi-disk reading) as shown in Figure 13.2. For example, a Secchi reading of 4 m intersects the lines for phytoplankton and attached plants near a depth of 10 m (4 × 2.5).

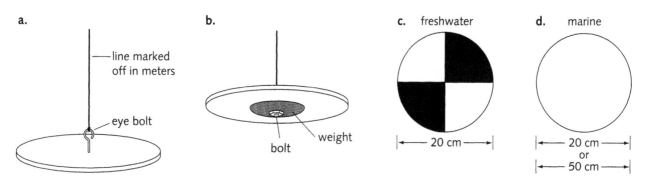

Figure 13.1   Secchi disk. (a and b) Construction. (c and d) Sizes and patterns of estuarine and marine disks.

Typical Secchi-disk readings for some water samples follow:

Distilled water	44 m
Caribbean Sea	41 m
Gulf of Maine	24 m
Chesapeake Bay (mouth)	3 m
Chesapeake Bay (mid-bay)	1 m

a. Lower the disk from the shaded side of the boat. Keep the line vertical (do not attempt to lower it when a strong current is flowing). The line should be marked in 0.1-m intervals.

b. Lower the disk until it just disappears from sight. Raise it until it just reappears. Measure this depth to the nearest 0.1 m by reading the mark at the water's surface.

c. Recoil the line and stow the disk in the proper place.

d. For each reading, consult Figure 13.2 to determine the approximate depth at which surface light intensity is reduced to the levels indicated by the four lines on the graph. Compare your readings with the typical readings mentioned earlier. You can also estimate photic depth.

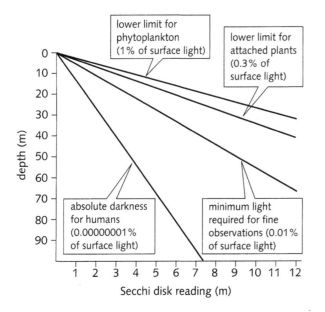

Figure 13.2 Approximate depth to which visible surface light can penetrate in water with Secchi-disk readings between 0 m and 12 m. Approximate depth of photic zone is indicated by the lines depicting the lower limits for phytoplankton and attached plants.

EXERCISE 2
Obtaining a Water Sample

There are a variety of ways to collect water samples for analysis of water quality. Simple devices, such as a bottle attached to a pole, can be used to sample water near the surface, but they cannot be used to obtain water from specific depths beneath the surface. Several kinds of devices are available for sampling water at specified depths, however. Some, like the **Nansen Bottle** (Fig. 13.3), are designed to sample deep water, and are equipped to carry reversing thermometers. Others, like the **Van Dorn sampler** (Fig. 13.4), are constructed of lighter-weight plastic.

All water samplers are designed as tubes that can be closed off at both ends at any selected depth. This design traps water inside the samplers and prevents contamination by other water or air as the devices return to the surface. Your instructor will explain how to use your sampler. It is essential that you understand how to transfer water from the sampler to containers for later analysis. You should also refer to methods described in Unit 1 (TSDO).

PROCEDURE

1. To transfer water from the sampling device to sample bottles, place the hose from the drain cock on the device in a sample bottle. Loosen the air vent on the device, and either open the drain cock or remove the hose clamp. If the sample will be analyzed for dissolved oxygen concentration, follow the steps below.

   a. Place the hose in the bottom of the sample bottle and allow the water to flow smoothly into the bottle. Gradually lower the bottle so that the hose remains beneath the surface of the water as it overflows.

   b. Close the drain cock or secure the pinch clamp, and carefully seal the sample bottle without trapping air.

   c. Chemically treat the sample for later oxygen analysis—refer to methods described in the TSDO exercise in Unit 1.

2. Dispense some of the water into a bucket thermometer (Fig. 13.5) for temperature readings.

3. Record the sample number and other essential information in your journal or on the oceanographic data log.

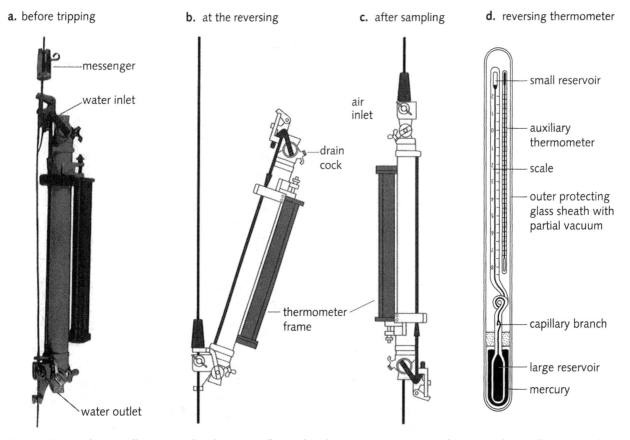

a. before tripping

messenger

water inlet

water outlet

b. at the reversing

drain cock

thermometer frame

c. after sampling

air inlet

d. reversing thermometer

small reservoir

auxiliary thermometer

scale

outer protecting glass sheath with partial vacuum

capillary branch

large reservoir

mercury

Figure 13.3   The metallic Nansen bottle is typically used in deep-water oceanographic research. (a) The water inlet and outlet are open upon descent through the water. (b and c) When a messenger trips the release, the bottle inverts its position by 180°. This action closes tapered valves at the water inlet and outlet. The inversion releases a second messenger for series sampling, and allows any number of Nansen bottles to be arrayed on one line. Each Nansen bottle can be equipped with a frame to hold one or more reversing thermometers (d). Mercury-filled thermometers are mounted in an inverted position. Inversion of the released Nansen bottle reverses the thermometer, and severs the mercury column at the constriction in the capillary branch. This action "locks in" the temperature of the water at the reversal depth.

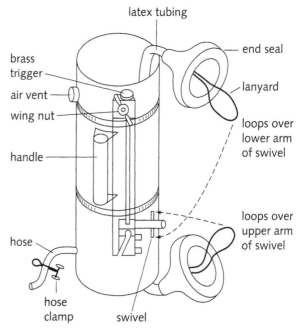

latex tubing

end seal

lanyard

loops over lower arm of swivel

loops over upper arm of swivel

brass trigger

air vent

wing nut

handle

hose

hose clamp

swivel

Figure 13.4   Van Dorn water samplers are routinely used in shallow-water research. The large-diameter acrylic or polyvinyl chloride cylinders offer little restriction to flow of water during descent, and can be used to sample both water and plankton. The samplers can be used in series. Plungerlike end seals are connected to each other by latex tubing, and are held open when their lanyards are looped over the arms of the T-shaped swivel. A messenger (small metal weight) sent down the line strikes the brass trigger, which rotates the swivel and releases the cups. They snap shut and close the sampler. If a second messenger was attached to the swivel, it is released and descends to the next sampler.

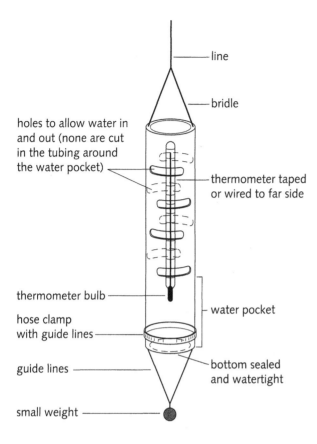

holes to allow water in and out (none are cut in the tubing around the water pocket)

line

bridle

thermometer taped or wired to far side

thermometer bulb

hose clamp with guide lines

guide lines

water pocket

bottom sealed and watertight

small weight

Figure 13.5 A simple model of a protected bucket thermometer.

EXERCISE 3

Measuring the Temperature of the Water Sample

Three methods for measuring water temperature are described.

PROCEDURE

1. For surface water samples, and in some cases for samples retrieved from relatively shallow depths, the temperature can be measured with a **bucket thermometer** (Fig. 13.5), a mercury thermometer suspended in a protective case with a reservoir at one end. Read the temperature as soon as the water sample is placed in the bucket.

2. A **thermistor** is a temperature-sensitive device connected to an electronic meter by a long cable marked off in meters. Dissolved oxygen meters and salinometers usually contain a thermistor on the probes.

   a. Unwind the line and lower the probe to the desired depth.

   b. Turn on the meter, calibrate it if necessary, and read (and record) the temperature on the dial.

3. If you are sampling water with a Nansen bottle (Fig. 13.3), you may be able to read the temperature from the **reversing thermometers** attached to the sampler (Fig. 13.3d). Your instructor will describe the mechanism and explain how readings are taken.

EXERCISE 4

Measuring the Salinity of the Water Sample

Several methods for measuring salinity are described here and in Unit 1 (TSDO).

PROCEDURE

1. A **salinometer** is an electronic device consisting of a probe connected to a meter by a cable. It can be lowered to any desired depth within limits of the cable to obtain a direct read-out of salinity, or electrical conductivity (and temperature—see Exercise 3).

2. Salinity can also be measured by chemical titration, with a hydrometer (density), or with a refractometer. These methods are described in Unit 1. Each method requires a discrete water sample from the given depth. Refer to directions given in Exercise 2 and Unit 1 for removing water samples from the sampling devices.

3. Record all observations and/or sample numbers in your journal or on the oceanographic data log.

EXERCISE 5

Measuring the Dissolved Oxygen Concentration of the Water Sample

Two methods for measuring dissolved oxygen (DO) concentration are mentioned here, and more fully described in Unit 1 (TSDO).

PROCEDURE

1. You can measure dissolved oxygen directly with a DO meter that contains a sensing probe connected to an electronic meter by a cable. The probe can be lowered to any desired depth, or it

can be inserted into a standard biochemical oxygen demand (BOD) bottle if water samples are retrieved.

2. The DO of water samples can be measured by chemical titration. In this case, it is essential that the water samples be properly collected and chemically treated (see Exercise 2 and the TSDO exercise in Unit 1 for details).

3. Record all observations and/or sample numbers in your journal or on the oceanographic data log.

B. Biological Sampling

Collecting organisms at sea is one of the more exciting ventures in marine biology. Although many coastal marine habitats have been sampled extensively enough to allow reliable prediction of species occurrence and distribution, surprises still occur. You never really know which organisms will appear on deck. If your trip will be your first encounter with this type of sampling, learn and enjoy!

A vast amount of gear has been designed for the capture and collection of marine organisms. The procedures that follow describe the use of four fundamental types of apparatus (gear) used to sample **plankton, demersal** (bottom-dwelling) **fish, epifaunal invertebrates,** and **benthic infauna.**

It is important to recognize that gear used to sample plankton, fish, and epifaunal invertebrates *concentrates* these organisms. Any tow samples substantial volumes of water and traverses large areas of sea bottom. The organisms captured in the collecting buckets were originally distributed throughout the water column or across the bottom. Accurate records of tow time, vessel speed, and area traversed enable you to estimate numerical abundance per unit volume of water or per unit area of sea bottom. While some variation is to be expected for all collection methods, benthic grabs and certain plankton tows are typically more reliable than epibenthic dredges and trawls.

While cruises are often undertaken to simply demonstrate the gear, projects can be designed to reveal differences in distribution and zonation of plankton, fish, and epibenthic organisms. For example, tows and grabs taken along the length of an estuary can demonstrate variations in biological communities along the salinity gradient. Plankton tows from different depths will show variation within the water column. Similar comparisons of channel and shoal water stations can be made.

EXERCISE 1
Making a Plankton Tow

Because most plankton are microscopic, they must be strained from the water with a funnel-shaped **net** (Fig. 13.6) of fine mesh (bolting silk or nylon—see Unit 8). The nets are towed behind a moving boat or held (tied to a bridge, for example) in a current. The plankton are funnelled into a **collection bucket** or bottle, from which the concentrated population can be retrieved for study.

PROCEDURE

1. Secure the net to a **tow line** from a deck winch, or to a separate line secured to a cleat near the stern of the boat.

2. Deploy the net while the boat is underway. **Do not** lower the net until you are instructed to do so. Lower the bucket end first, paying out the tow line slowly as the net moves away from the boat. If the sample is to be taken at some depth beneath the surface, a **depressor** weight may be attached to the towing line in front of the net.

3. Record the following information in your journal or on plankton report forms (see Unit 8): date, location (station number), tow sample number, diameter of net, mesh size of net, start time of tow, and vessel or current speed. Write the same information on a sample label for the collection bottle.

   Note the slow speed of the vessel. Why is the boat traveling so slowly?

4. At the end of the tow, haul in the net. Keep the mouth of the net up, and the bucket end down. Record the end time of tow in your notebook.

5. Gently wash down the outside of the net with seawater. This action flushes any plankton stuck in the netting into the bucket.

6. Remove the plankton bucket (or open its end) and empty the plankton into a collection bottle (if the sample is to be preserved) or a bucket of seawater. The collection may be preserved by adding sufficient buffered formalin to the sample to achieve a 5% solution.

7. Analyze the sample as described in Unit 8.

EXERCISE 2

Using an Otter Trawl to Sample Fish and Invertebrates

Organisms that live on or near the sea bottom that is relatively level and free of obstructions can be sampled with a trawl net similar to those used in many commercial fishing operations. The otter trawl is funnel-shaped—that is, wide at the mouth and narrow at the **cod end** (where the catch is trapped) (Fig. 13.7) Floats on the **headline** and weights (often a chain) on the **foot rope** keep the mouth of the net open and on the bottom. Weighted panels, called **otter boards** or **doors,** are positioned ahead of the net. These panels spread the net as the boat pulls it forward. A **trailing buoy** is often attached to the cod end of the net. In the event of a snag, the buoy can be retrieved to free the net.

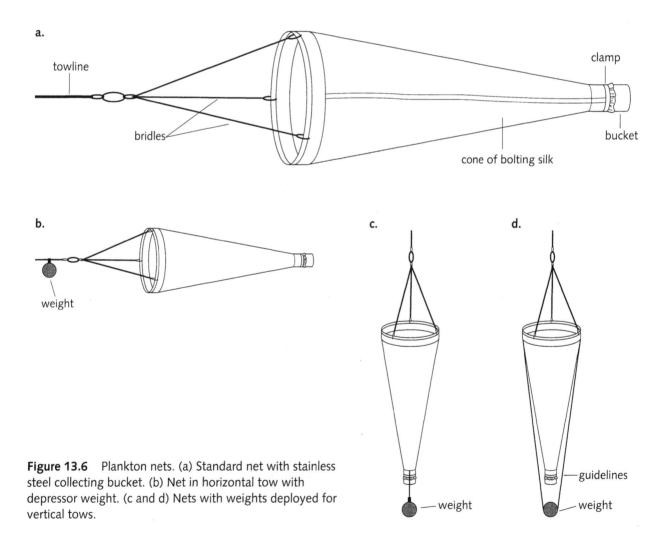

Figure 13.6 Plankton nets. (a) Standard net with stainless steel collecting bucket. (b) Net in horizontal tow with depressor weight. (c and d) Nets with weights deployed for vertical tows.

PROCEDURE

1. Determine (and record on the trawl collection data log) the depth of the water at the study site. Depth of water determines the length of the tow line, which is usually four times longer than the depth.

2. Follow the instructions of the captain or crew, or watch carefully.

3. Make sure that the cod end of the net is tied shut.

4. Check that chains and lines are not crossed or tangled.

5. When the captain gives the order to set the net, place the trailing buoy in the water at the stern. The forward movement of the boat will allow you to pay out the line. Keep your feet clear of lines on the deck!

6. Lower the rest of the net off the stern, cod end first.

7. Carefully allow the doors to slide off the stern and verify that they begin to spread apart.

8. Pay out the tow line at a steady rate until the desired length of line is deployed. Set the winch, and record the start time of tow, and vessel speed (why?).

9. To retrieve the net, reverse the operations described above. Record the end time of tow (why?). Pull the doors into the boat, and place them forward and out of the afterdeck area. Lift and shake the net as it comes aboard to force the catch into the cod end. Pull in the trailing buoy and line, and secure them on deck.

10. Hold the cod end over a culling table or a large container. Untie the line securing the end of the cod end and release the catch.

11. Sort the catch into different containers.

CAUTION

Use gloves during this operation to protect yourself from fish spines and other potential hazards such as jellyfish. Beware of flapping fish in a catch containing jellyfish. Detached pieces of tentacles contain active stinging cells that are particularly irritating to the face.

12. Identify and sort the catch. Record numbers of individuals, sizes, and wet weights, in your journal, or use the trawl collection data logs and the size/weight—size frequency logs.

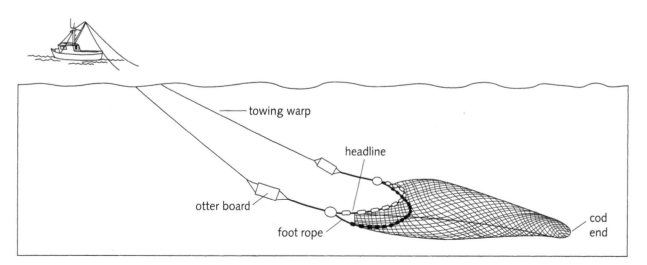

Figure 13.7 An otter trawl deployed from a fishing trawler. Otter boards spread the mouth of the net, which has a weighted foot rope and headline with floats. Fish are swept into cod end, which usually is made of smaller mesh netting than the body of the net.

EXERCISE 3

Sampling the Epifauna with a Biological Dredge

A biological dredge (Fig. 13.8) is designed to sample organisms living in, or directly above, the sea bottom. It is often used on bottoms covered by rocks or heavy shell debris. The procedure for its use resembles that of the otter trawl, with the principal difference involving the manner of emptying the catch from the bag. Because the end of the bag has no opening, the catch must be dumped out through the mouth of the dredge. Record your observations on the dredge collection data log.

EXERCISE 4

Sampling the Benthic Infauna with a Bottom Grab

A wide assortment of devices are used to quantitatively sample the organisms that live in the sea bottom. Bottom grabs (Fig. 13.9) are deployed in free-fall from the research vessel so that they embed themselves into the sea bottom. A triggering device then permits closure of the jaws prior to, or during, retrieval. The area of bottom sampled is equivalent to the area of the mouth of the grab when the jaws are fully opened. The following procedure describes the operation of the **Petersen grab.**

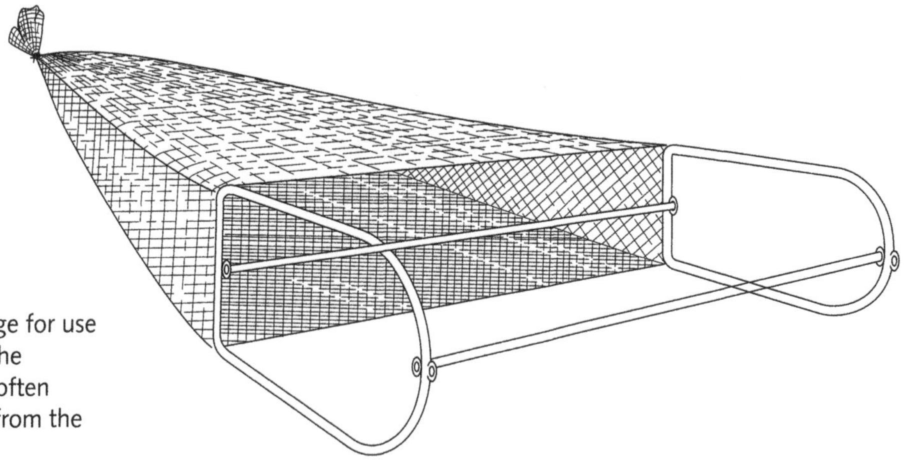

Figure 13.8 A biological dredge for use on irregular, rugged bottoms. The bottom portion of the frame is often toothed to dislodge organisms from the bottom.

Figure 13.9 A Petersen bottom grab. A steel bar keeps the jaws locked open until it hits the bottom. The safety pin prevents accidental closing when the open grab is handled on deck.

Grabs are heavy and potentially dangerous. Always keep your fingers away from the jaws and closing mechanisms. The safety pin must be in place whenever the grab is lifted in the open position. Pay attention to instructions.

PROCEDURE

1. With the grab on the deck, open the jaws and insert the safety pin (Fig. 13.9).

2. Raise the grab by hauling in the winch cable, and swing the grab overboard. Hold the grab firmly on the outside of the jaws if the boat is rocking.

3. When directed to do so, remove the safety pin and slowly lower the grab to just below the surface of the water.

4. When directed, release the lock on the winch so that the grab drops freely to the bottom.

5. Give the line a few quick jerks to ensure that the locking bar releases the jaws. Retrieve the grab.

6. Swing the grab back on board and lower it into a sampling container or screen.

7. Open the jaws and gently wash the sediment into the container or screen.

8. Describe the sediment or save a small jar of it for meiofaunal and/or sediment analysis (see instructions in Units 4 and 12).

9. Gently wash the sediment through the sorting screen. Transfer the organisms collected on the screen to jars that have been appropriately labeled (date, station, depth) and filled with seawater (for examination of living forms) or 5% formalin in seawater.

10. Complete the analysis of the samples in the laboratory, and record your observations on the infauna analysis data log or in your journal.

OCEANOGRAPHIC DATA LOG

Date _____   Vessel _____

SURFACE DATA

| Station No. | Location | Time (h) | Baro. Press. (mbs) | Air Temp. | | Wind | | |
				Dry (°C)	Wet (°C)	Direction	Speed (mph)	Beaufort Scale
___	___	___	___	___	___	___	___	___
___	___	___	___	___	___	___	___	___
___	___	___	___	___	___	___	___	___
___	___	___	___	___	___	___	___	___
___	___	___	___	___	___	___	___	___

| Wave | | | | | | | | | Other |
Direction	Height (m)	Period (sec)	Sea State	Light (ft-cdl)	Tide Stage	Visibility (km)	Cloud Type	Weather	% Cloud Cover	
___	___	___	___	___	___	___	___	___	___	___
___	___	___	___	___	___	___	___	___	___	___
___	___	___	___	___	___	___	___	___	___	___
___	___	___	___	___	___	___	___	___	___	___
___	___	___	___	___	___	___	___	___	___	___

OCEANOGRAPHIC DATA LOG

SUBSURFACE DATA

Station No.	Time (h)	Depth (m)	Water Temp. Therm. (°C)	Water Temp. Meter	Light Secchi (m)	Light Photo. (% T)	Light Trans. (% T)	Current Speed	Current Direction	Density (g/cm³)
___	___	___	___	___	___	___	___	___	___	___
___	___	___	___	___	___	___	___	___	___	___
___	___	___	___	___	___	___	___	___	___	___
___	___	___	___	___	___	___	___	___	___	___
___	___	___	___	___	___	___	___	___	___	___
___	___	___	___	___	___	___	___	___	___	___
___	___	___	___	___	___	___	___	___	___	___
___	___	___	___	___	___	___	___	___	___	___
___	___	___	___	___	___	___	___	___	___	___
___	___	___	___	___	___	___	___	___	___	___
___	___	___	___	___	___	___	___	___	___	___
___	___	___	___	___	___	___	___	___	___	___

Salinity (o/oo) Cond. Meter	Salinity (o/oo) Chem.	Salinity (o/oo) From Dens.	O₂ (ppm) Meter	O₂ (ppm) Chem.	% Sat	pH	NO₃-N (ppm)	PO₄-P (ppm)	Water Sample Bottle Number	Water Sample Station No.
___	___	___	___	___	___	___	___	___	___	___
___	___	___	___	___	___	___	___	___	___	___
___	___	___	___	___	___	___	___	___	___	___
___	___	___	___	___	___	___	___	___	___	___
___	___	___	___	___	___	___	___	___	___	___
___	___	___	___	___	___	___	___	___	___	___
___	___	___	___	___	___	___	___	___	___	___
___	___	___	___	___	___	___	___	___	___	___
___	___	___	___	___	___	___	___	___	___	___

OCEANOGRAPHIC DATA LOG

BOTTOM SUBSTRATE

| Station No. | Depth to Bottom | | Type of Sampler | Sample Color | Sample Odor |
	Sound. Line (m)	Depth Sounder (m)			
_____	_____	_____	_____	_____	_____
_____	_____	_____	_____	_____	_____
_____	_____	_____	_____	_____	_____
_____	_____	_____	_____	_____	_____
_____	_____	_____	_____	_____	_____

Sediment Type Field Description	Sample Color	Sediment Mineralogy	Other
_____	_____	_____	_____
_____	_____	_____	_____
_____	_____	_____	_____
_____	_____	_____	_____
_____	_____	_____	_____

NAME _____ SECTION _____ DATE _____

TRAWL COLLECTION DATA LOG

Collection No. _____ Station No. _____

Vessel _____ Date _____ Time _____

Depth of Water (m) _____ Vessel Speed _____

Set Time _____ End Time _____ Duration of Tow _____

Trawl Net: Width of Mouth (m) _____

Mesh Size in Cod End _____

Area of Bottom Sampled (m²) _____

Bottom Sediments _____

Fish Species	Number/ Weight/ Volume of Sample (/m²)	Invertebrate Species	Number/ Weight/ Volume of Sample (/m²)
_____	_____	_____	_____
_____	_____	_____	_____
_____	_____	_____	_____
_____	_____	_____	_____
_____	_____	_____	_____
_____	_____	_____	_____
_____	_____	_____	_____
_____	_____	_____	_____
_____	_____	_____	_____
_____	_____	_____	_____
_____	_____	_____	_____

NAME _____ SECTION _____ DATE _____

DREDGE COLLECTION DATA LOG

Collection No. _____ Station No. _____

Vessel _____ Date _____ Time _____

Depth of Water (m) _____ Vessel Speed _____

Set Time _____ End Time _____ Duration of Tow _____

Gear: Width of Mouth (m) _____

           Depth of Sample _____

Area of Bottom Sampled (m²) _____

Bottom Sediments _____

Fish Species	Number/ Weight/ Volume of Sample (/m²)	Invertebrate Species	Number/ Weight/ Volume of Sample (/m²)
_____	_____	_____	_____
_____	_____	_____	_____
_____	_____	_____	_____
_____	_____	_____	_____
_____	_____	_____	_____
_____	_____	_____	_____
_____	_____	_____	_____
_____	_____	_____	_____
_____	_____	_____	_____
_____	_____	_____	_____

BOTTOM GRAB COLLECTION DATA LOG
INFAUNA ANALYSIS

Collection No. _____ Station No. _____

Vessel _____ Date _____ Time _____

Depth of Water (m) _____

Bottom Sampler (m) _____ Sample Size (m²) _____

Bottom Sediments _____ Sieve Size _____

Taxon	No. of Individuals in Sample (/m²)	Taxon	No. of Individuals in Sample (/m²)
_____	_____	_____	_____
_____	_____	_____	_____
_____	_____	_____	_____
_____	_____	_____	_____
_____	_____	_____	_____
_____	_____	_____	_____
_____	_____	_____	_____
_____	_____	_____	_____
_____	_____	_____	_____
_____	_____	_____	_____
_____	_____	_____	_____

NAME _____ SECTION _____ DATE _____

SIZE/WEIGHT OR SIZE FREQUENCY SHEET

Species _____

Collection No. _____ Station No. _____

Location _____ Date _____ Time _____

Specify dimensions and units of measurements _____

Size ( )	Count	Weight (g)	Size ( )	Count	Weight (g)
———	———	———	———	———	———
———	———	———	———	———	———
———	———	———	———	———	———
———	———	———	———	———	———
———	———	———	———	———	———
———	———	———	———	———	———
———	———	———	———	———	———
———	———	———	———	———	———
———	———	———	———	———	———
———	———	———	———	———	———
———	———	———	———	———	———
———	———	———	———	———	———
———	———	———	———	———	———
———	———	———	———	———	———
———	———	———	———	———	———
———	———	———	———	———	———
———	———	———	———	———	———
———	———	———	———	———	———
———	———	———	———	———	———
———	———	———	———	———	———
———	———	———	———	———	———
———	———	———	———	———	———
———	———	———	———	———	———
———	———	———	———	———	———
———	———	———	———	———	———
———	———	———	———	———	———
———	———	———	———	———	———

Microscopes in the Marine Biology Laboratory

This appendix contains guidelines for the care and handling of microscopes, descriptions of compound and binocular dissecting microscopes, and instructions for measuring the size of microscopic objects.

RULES FOR THE CARE AND HANDLING OF MICROSCOPES

- When you transport any microscope, hold it upright with one hand on the arm and the other under the base. Do not carry any other item while you are transporting the microscope. Do not jar the instrument when you place it on the table.

- Clean the lenses with grit-free lens paper before and after use.

- Clean the other components with lab toweling. Seawater is corrosive; wipe up any spills of seawater immediately, clean the components with a moist sponge, and dry them.

- Never remove any parts from the microscope unless you are instructed to do so.

- Inform the instructor of any mechanical problems.

- Always begin the focusing process with the lowest-power objective lens in position.

- Prior to storage, remove any slides or specimens from the stage, return the lowest-power objective lens to the start position, and replace the dust cover.

THE COMPOUND MICROSCOPE

Although all compound microscopes contain similar optical systems, their parts are arranged differently. Become familiar with the model in your laboratory, by comparing it with Figure A1.1, identifying the following parts, and understanding their function.

The **base** supports the microscope and houses the **substage light** or **in-base illuminator.** This light may be controlled by a simple on–off toggle switch (light intensity is fixed) or by an additional control that allows you to vary the intensity.

The **arm** is the vertical component that supports the **head.** The head contains the **ocular** lens (eyepiece), through which you make your observations. The **nosepiece** extends from the lower part of the head and carries two to four **objective lenses.** The nosepiece may be rotated to bring each lens into position and thereby select different magnifying powers.

The **coarse** and **fine adjustment knobs** are found near the bottom of the arm. The coarse adjustment is used with the low-power objectives (scanning lens = 4× and 10×) **only!** The fine adjustment may be used with any of the objectives.

The **stage** is the platform upon which the slide is placed for observation. Some stages have simple **stage clips** to accomplish this purpose; others have an accessory **mechanical stage** that holds the slide. The mechanical stage permits precise movement of the specimen and is quite useful for taking counts and measurements. If you do not know how to secure your slide on your stage, ask your instructor.

The **condenser lens** and **iris diaphragm** are located beneath the stage, in direct alignment with the light source. The condenser lens focuses the light on the specimen, and *should be left in the full upright position* for ideal viewing. It is lowered only when the lens requires cleaning. The best image (contrast) is obtained by adjusting the light intensity. Contrast is enhanced, not by increasing the intensity, but by decreasing it. Move the iris diaphragm (by the lever extending outward from the condenser) to control the amount of light passing through the condenser lens.

Examine the objectives on your instrument. Ordinarily the lowest-power lens has a magnification

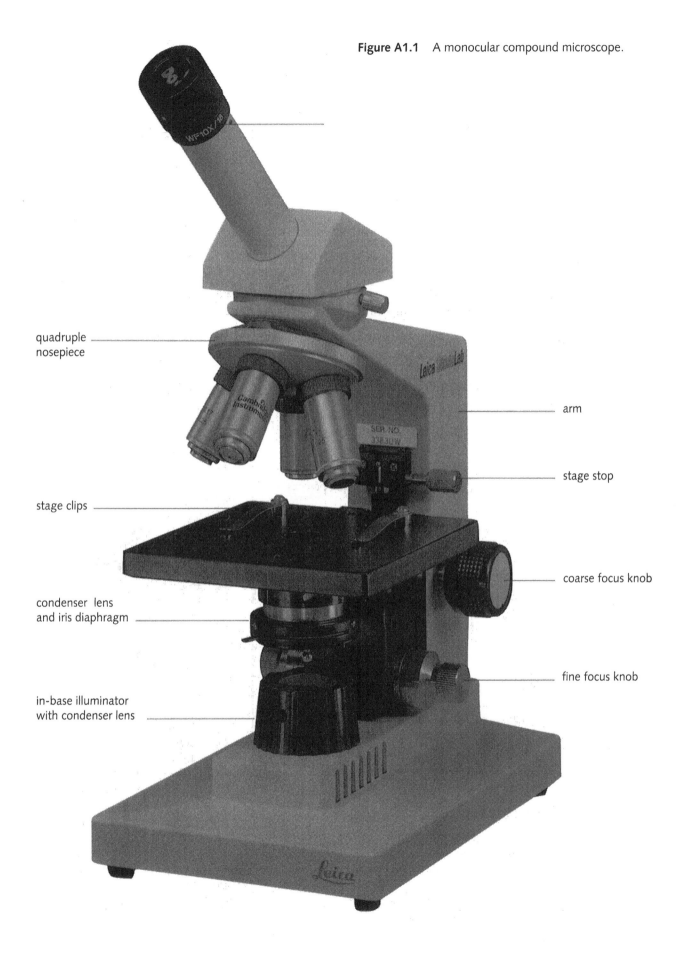

Figure A1.1 A monocular compound microscope.

quadruple nosepiece

stage clips

condenser lens and iris diaphragm

in-base illuminator with condenser lens

arm

stage stop

coarse focus knob

fine focus knob

Microscopes in the Marine Biology Laboratory

of 10× (the number is inscribed on the side of the objective). You also may have a **scanning lens** of 4× magnification. If so, it is about half the length of the 10× objective. The **high-dry** objective has a magnification from 40× to 45×, and is longer than the 10× lens. The **oil immersion** objective is even longer, with a magnifying power of 95× to 100×. To use this objective, place a drop of oil between it and the coverslip of the slide to seal the space. This procedure must be carefully performed, and will be demonstrated by your instructor.

Each objective is associated with a specific **working distance**—the distance between the bottom of the objective lens and the top of the coverslip. The distance can be large (several millimeters) for the scanning lens or extremely small (less than 1 mm) for the oil immersion lens. The small working distance is the reason you use only the fine adjustment when using the high-dry or oil immersion objectives.

Magnification of the image is achieved by the combination of the ocular and objective lenses. The **total magnification** of any specimen under observation is equal to the product of the magnifying powers of the two lenses in position. For example, a 10× ocular and an objective with 43× magnification results in an image 430 times greater than its actual size. Be certain to express the total magnification when you make drawings of microscopic specimens. You can then estimate the actual size of the organism represented by your drawing. A more accurate way to provide this information is to measure the specimen (see Section D later in this appendix).

THE DISSECTING MICROSCOPE

The dissecting microscope allows you to observe specimens too large to be viewed with the compound microscope. Many different kinds of dissecting microscopes exist, from fixed to variable magnification models, with or without a variety of accessories. They are versatile and extremely useful in marine biology. Compare your instrument with the following description and Figure A1.2.

The **base** contains a large opening that holds two types of **plates.** A reversible metal plate (black on one side, white on the other) is normally employed when the **trans-illumination base** is not used, and when either the opaque black or white background is desirable for maximum contrast with the specimen. A **glass** plate is used concurrently with trans-illumination. Note that the trans-illumination base contains an **adjustable mirror** (with two surfaces: plano diffusing and concave reflecting) and an

opening **(port)** on the rear of the base near the arm. The port supports a Nicholas illuminator.

The arm contains one set of **focusing knobs** (but no fine focusing adjustment) and supports the **power pod,** which contains the **magnification knob** and the **oculars.** Note that one of the oculars contains a knurled **focusing ring** at its base. The position of the power pod may be reversed by swinging out the two **lock levers** at either side of the supporting arm. The supporting arm for the power pod also contains a port for an illuminator.

To become acquainted with your instrument, place a simple object on the stage. Adjust the oculars to the proper distance between the pupils of your eyes so that both fields are viewed as one. This action allows you to view a three-dimensional image.

Set the magnification knob to the highest power (2× to 3×). Now look through the right ocular, using only your right eye (close your left eye) and focus on the object. Adjust the focusing knob until the image is sharp. Readjust the magnification knob to the lowest power (0.7× to 1.0×). Without moving the focusing knob, look through the left ocular and (using only the left eye) adjust the ocular focusing ring, clockwise or counter-clockwise, until the image comes into sharp focus. If you follow this procedure carefully, you can set the magnification at any value within the range of the power pod without having to refocus. Plastic **eyeguards** exclude stray light and orient the eyes in the proper relationship to the eyepieces. Eyeguards should not be used if you wear eyeglasses.

Variable focusing microscopes allow you to change the power continuously to obtain the best magnification for a given specimen. The magnification knob is engraved in steps of 0.1× throughout its power range. Total magnification is determined by multiplying the reading on the magnification knob by that of the oculars (available in 10×, 15×, and 20×). If your instrument has an **accessory lens** (available in 2× and 1/2× magnifications), it will be attached to the bottom of the power pod.

Opaque specimens may be illuminated in several ways, depending on the nature of the specimen's surface. Specimens may be placed on the glass stage plate, or if a contrasting background is desired, on the black and white plate. If the surface is light-diffusing (for example, a clam shell), use oblique lighting such as a Nicholas illuminator, which can be inserted in the port on the power pod support. A fluorescent illuminator can be used to provide cool, diffuse light.

The glass stage plate must be used with trans-illumination for transparent or semi-transparent specimens. Insert a Nicholas illuminator in the port at the base or the stage, or focus it on the substage mirror. The mirror may be used in conjunction with any available source of light. Where high brightness is needed, use the concave reflector side. When diffuse lighting is required, use the plano diffusing surface. The concave reflector also may be employed to produce oblique lighting.

DETERMINING THE SIZE OF MICROSCOPIC OBJECTS

The size of microscopic objects can be accurately measured with an **ocular micrometer,** a scale that has been etched in a glass disk and inserted within the ocular of the microscope. You can also estimate the size of an object, with a lesser degree of accuracy, by comparing the object's size to that of the di-

ameter of the microscopic field. Directions for both methods, which are applicable to both compound and dissecting microscopes, follow.

Calibration of Ocular Micrometers

You will need two scales: the **ocular micrometer,** a glass disk bearing an *arbitrary scale* of 50 to 100 divisions (upper scale shown in Fig. A1.3), and a **stage micrometer,** a glass slide etched with a *known scale* of 1 mm or 2 mm usually subdivided into units of 0.1 mm and 0.01 mm (Fig. A1.3, lower scale). Because the ocular micrometer scale is arbitrary, the scale value differs for each objective, and therefore must be calibrated against a standard scale on the slide micrometer.

Place the stage micrometer on the microscope stage and bring it into sharp focus at the lowest-

(a)

(b)

Figure A1.2   (a) Binocular dissecting microscopes. Model in rear is mounted on weighted base with adjustable extension arm for viewing large specimens. Model in front is on conventional stand; stage is equipped with clips and a metal disk with white side up. Other models have a deeper stage and glass disk so that indirect illumination via reflected light may be used. (b) Nicholas illuminator. The lamp may be used as shown, or it may be connected to a dissecting microscope. The tapered barrel is designed to fit into a tapered hole behind the head of the microscope or in the substage unit.

power magnification. Adjust the ocular and stage micrometer scales so that they are parallel and their zero lines coincide, as shown in Figure A1.3. Locate the point toward the right where lines from the two scales also coincide. Count the number of lines on the stage micrometer (SM) and the number of lines on the ocular micrometer (OM) between the zero point and the point where the lines coincide exactly. Then divide the number of SM lines by the number of OM lines. Because the number of SM lines represents a real value (in millimeters), this division reveals the value of one OM unit (OMU) for that specific objective lens. Repeat the procedure for all objective lenses.

In the example shown in Figure A1.3, 70 divisions on the OM scale are equal to 24 divisions (= 0.24 mm) on the SM. The division SM/OM = value of each OMU, or

0.24 mm/70 = 0.0034 mm

Since there are 1000 μm in 1 mm, then 1 OMU = 3.4 μm (= 0.0034 mm × 1000). Two OMU = 6.8 μm, 5 OMU = 17 μm, and so on.

Determining the Size of the Microscopic Field

If an ocular micrometer is unavailable, you can calibrate the diameter of the microscopic field and the length/width of the pointer in the ocular. Using the stage micrometer as a measuring rule, you can measure field diameter or the pointer. Be certain that you know the units and length of the stage micrometer.

If you do not have access to a stage micrometer, you can calibrate the microscope field in the following manner, which is particularly useful with dissecting microscopes. Obtain a microscope slide covered with graph paper ruled in millimeters. Each square in the grid is 1 mm on each side. Focus in low power.

Move the slide so that one grid line touches the edge of the field on one side. Count the number of squares across the diameter of the field. If you see only part of a square, estimate the part of a millimeter that the partial square represents. Record this figure in your notebook or laboratory manual.

The field diameter visible through the other objectives is proportionally smaller than the field diameter that you just measured. It also will be very difficult to obtain suitable light transmission through the paper at higher magnifications on the compound microscope. It is possible, however, to compute the field diameter of the other magnification fields by using the formula:

(Diameter of the low-power field) × (Total magnification of the low-power field) = (Diameter of field Y) × (Total magnification of field Y)

For example, say the diameter of the 40× field is 2 mm. You can compute the diameter of the 100× field as:

2 mm × 40 = Y (diameter of 100X field) × 100, or
(2 mm × 40) = 100 Y, or
80 mm = 100 Y, or
0.8 mm = Y (diameter of 100X field), or 800 μm

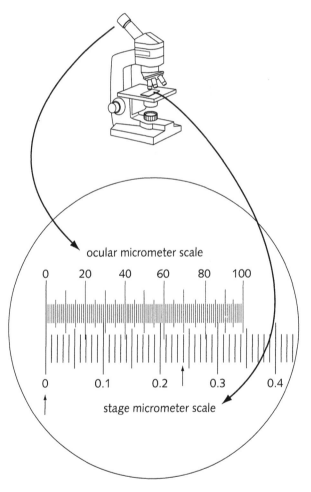

Figure A1.3 Alignment of ocular micrometer scale (top) with the stage micrometer scale (bottom). In this example, the two scales are in perfect alignment at the zero line (on the far left) and the 70 mark of the ocular micrometer (designated by arrows). The 70 mark coincides with the 0.24 mm mark of the stage micrometer.

How to Analyze and Present Data

By the time you finish any study in which you make quantitative observations, you have accumulated a substantial amount of "raw" data. You need to convert your data (by *data analysis*) into some simplified form that is understandable, not only to you, but also to your peers. This activity, called "number crunching" by investigators, involves more than mathematical calculations. Numerical data must be *presented* in meaningful ways that aid you and the reader in recognizing and understanding trends, and inter-relationships that might be hidden in your raw data. Several types of numerical computations and graphing techniques fundamental to data analysis are described in this appendix.

DATA ANALYSIS

This section contains three parts. The first section covers **descriptive statistics:** mean, median, mode, standard deviation, standard error, and range. These measures are applicable to nearly every marine biology course.

The second section briefly introduces **statistical analysis,** a highly effective tool used to determine mathematically if data sets are statistically similar or different. Such analyses may be used in more advanced marine biology courses.

The third section describes mathematical applications that lead to calculation of population parameters: **diversity** and **relative importance indices.** Although these calculations may appear complex, in fact they are not. They are particularly useful in determining differences between communities, and in assessing the relative importance of individuals in any given community.

Descriptive Statistics

Summary statistics, such as those listed in Table A2.1, represent a simplification of a "raw" data set.

They make the data more manageable and facilitate plotting graphs. A few calculations are described below.

1. **Arithmetic mean** ($\overline{X}$) is calculated by adding the measurements (ΣX) and dividing by the number (n) of measurements (observations):

   $X = \Sigma X/n$, or
   $X = (X_1 + X_2 + \ldots + X_n)/n$

2. **Standard deviation** (s, σ, or SD), and **Standard error of the mean** (SE) describe the distribution of a population about its mean. The SD measures the distance of frequency distribution of observations (n) from the mean. One SD on either side of the mean includes 68% of the

Table A2.1 Descriptive statistics for dry weight and total glycogen content of ovaries from 21 shrimp. These summarized data correspond to the scatterplot and regression line shown in Figure A2.1.

Parameter	Weight (mg)	Glycogen Content (μg)
Number of observations (*n*)	18	18
Mean ($\overline{X}$)	48.20	86.68
Standard deviation (SD)[*]	+/−22.38	+/−58.48
Standard error (SE)[*]	+/−5.28	+/−13.78
Range	93.95	164.32
Minimum	22.55	21.18
Maximum	116.50	185.50
Median	44.57	61.73

[*]The +/− prefixes to the SD and SE values in this table serve as a reminder that the values extend on both sides of the mean, by definition. It is not necessary (and often discouraged) to include the +/− signs in such data presentations.

possible data values; 2 SD = 95%. The SD is easily derived:

$$SD = \Sigma \, X^2 - [(\Sigma X)^2/n]/n - 1$$

where $\Sigma \, X^2$ = sum of the square of each data point; $(\Sigma X)^2$ = square of the sum of all data points; and n = number of observations. Reporting both SE and SD is unnecessary because one can be derived from the other:

$$SE = SD/\sqrt{n}$$

3. **Range** is the spread between the smallest **(minimum)** and largest **(maximum)** values in the data. Presenting both minimum and maximum values is more accurate than reporting a single value for range.
4. **Median** is the single measurement that represents the midpoint of the data set. Half the measurements are larger than the median; half are smaller.
5. **Mode** is the value that occurs most frequently in a list of measurements, and can be recognized in a frequency distribution plot. In Figure A2.7, for example, the mode is the 20-mg weight class of the displayed population.

Statistical Analysis

In the biological world, variability is commonplace. Consider the following: No two organisms respond the same way to any given environmental condition; the numbers and sizes of individuals of a particular species vary in time and space; unavoidable errors in observations and measurements compound the problem.

Fortunately, most of the variability you record reflects real biological differences among the individuals in the sample population. You should not ignore such variability in your data presentation and interpretation.

Although some variability is detectable in descriptive statistics, you should determine whether observed variability within a sample (or differences between two or more sample populations) is biologically significant or results from random chance.

Statistical tests, such as chi-square, and student-t, determine the probability that the observations were due to chance alone. If you are asked to perform a statistical analysis of your data, your instructor will specify the test, and guide you through the necessary steps, from establishing the null hypothesis to interpreting the derived statistical value.

Regression analysis is a statistical method used to describe mathematically the relationship between two variables, often illustrated in a graph

(Fig. A2.1). The regression equation enables you to estimate the value for the dependent variable (y) for any given value of the independent variable (x). The regression coefficient (r) indicates how well the plotted line represents the observed data.

Population Parameters

Numerous mathematical procedures have been designed to describe quantitatively the composition of populations and communities. Several are described here.

1. **Species Diversity.** Different communities support different numbers and species of organisms. It is generally believed that, as the number of species increases in a community, the stability of that community increases. Species diversity tends to be low in physically controlled ecosystems and high in biologically controlled systems.

   The simplest measure of species diversity involves a count of the numbers and kinds of organisms existing in a marine community.

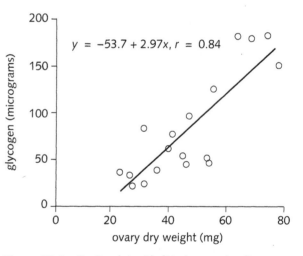

Figure A2.1 Scatterplot with fitted regression line. Ovary dry weight, on the x (abscissal) axis, is the independent variable; glycogen content of the ovary, on the y (ordinal) axis, is the dependent variable. Each open circle represents a single data point that indicates the glycogen content of an ovary at that particular weight. The solid line running through the data points represents the equation fitted to the data: y = glycogen content, x = ovary weight, 2.97 = slope of the line, −53.7 = y intercept (the location along the y axis where the line would intersect if extended downward to that point), r = regression coefficient that indicates how well (or poorly) the fitted line describes the data. In this case the fitted line accounts for 70.6% (r^2) of the variation.

Species counts alone do not, however, account for two important factors: abundance patterns and sample size. For example, two communities might contain 4 species each and 100 individuals. The first community might consist of 97 individuals of one species and one individual of each of the other 3 species. The second community might include 25 individuals of each species. The second community is more diverse than the first.

Diversity has two components: (a) richness, the number of species present; and (b) evenness, a measure of the distribution of population sizes of the respective species. Evenness approaches its maximum as the population sizes of the species approach equality.

Several diversity indices allow an assessment of various sampling problems. Two commonly acceptable indices are the **Shannon-Wiener** index (H) and **Simpson's** index (D). The formulae for these indices are

$$H = - \Sigma \, (n_i/N_s) \ln \, (n_i/N_s)$$
$$D = 1 - \Sigma \, (n_i/N_s)^2$$

where n_i = the number of individuals belonging to each (the ith) species; N_s = the total number of individuals belonging to all species in the sample (station).

Note that $\log_2$, $\log_{10}$, or natural log (= ln) can be used in the Shannon-Wiener function. The units of diversity using this index depend on the base of the logs—bits, decits, or nats, respectively. Indices using $\log_{10}$ or ln can be converted to $\log_2$ as follows:

$$\log_2 X = 1.4427 \ln X$$
$$\log_2 X = 3.3219 \log_{10} X$$

The combination of two measures—the number of species in the sample and of the relative abundance patterns (H or D)—summarizes information on biological diversity. The Shannon-Wiener index approaches maximum when the proportion (p) of the species in the community is nearly equal: $p_1 = p_2 = \ldots = p_n$. Simpson's measure approaches a maximum of 1.00.

The manipulation of real data illustrates the use of diversity indices more accurately. Begin by examining the sample field data in Table A2.2. Four sites (stations) were sampled. A total of 457 individual organisms, representing 16 species, were collected. Of the 16 species, 11, 6, 9, and 7 were present at stations 1–4, respectively.

Now examine Table A2.3. The top portion presents a breakdown of calculations required to determine the diversity indices for the Lam-

oine Pool station, which contained 163 organisms representing 11 species. The lower three parts of the table list the species from the other stations and their diversity indices.

When these indices are plotted in relation to number of species (Fig. A2.2), you can see that diversity indices increase in direct proportion to species number.

2. **Index of Similarity**. If you obtain samples of a certain community at several locations, such as depicted in Table A2.2, you also may be interested in objectively determining which stations are similar to each other and which are dissimilar. A simple index (SI) can be calculated

$$SI = 2C/(A + B)$$

where A = number of species at Station A; B = number at Station B; and C = number of species common to both. After SI has been computed for all combinations, the values are placed in a matrix that makes comparisons among the stations easy.

Consider the sample field data (Table A2.4). In this example, Sandy Point is very similar in composition to Lamoine Pool. Fort Point is least similar to Lamoine Jetty.

3. **Index of Relative Importance (IRI).** Combinations of two or three separate measurements (frequency, number, volume, or mass) of species occurrence in a community (or of prey items in a gut analysis study) may reveal more about the situation than each parameter evaluated individually. The IRI usually combines three measurements and allows you to rank the species

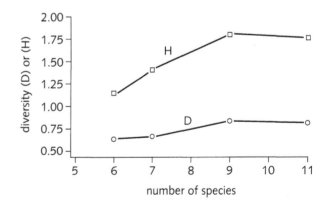

Figure A2.2 Line graph illustrating the relationship between computed diversity indices and the number of species at four stations (from Table A2.2). Computations of the indices (H = Shannon-Wiener index, D = Simpson's) are shown in Table A2.3. The two indices are plotted with different symbols and lines, and are directly labeled in the field of the graph.

by the combined index and display them graphically (Fig. A2.3). In addition, meaningful information can sometimes be obtained with two parameters.

The three parameters most commonly used are *numerical composition (NC)*, *gravimetric composition (GC)*, and *frequency of occurrence (FO)*. These measures are expressed as percentages. For example, numerical composition is calculated

$$NC = N_i/N_t \times 100$$

where N_i is the total number of individuals of each species (*i*) ($N_i = \Sigma\ n_i$), and N_t is the total number of individuals of all species present in all samples.

Gravimetric composition is calculated

$$GC = M_i/M_t \times 100$$

where M_i is the total mass (weight or volume) of individuals of each species (*i*) ($M_i = \Sigma\ m_i$), and M_t is the total mass of all species present in all samples.

Frequency of occurrence is calculated

$$FO = (\Sigma s_o)/S_t \times 100$$

where s_o is the number of stations at which species (*i*) occurs, and S_t is the total number of all stations sampled.

As an example of the determination of IRI, we again refer to the sample field data (Table

Table A2.2 Number (n_i) and biomass (m_i) of intertidal macrofaunal invertebrates collected at four stations on the Maine coast.

TAXA	Lamoine Pool		Lamoine Jetty		Sandy Point		Fort Point		SPECIES TOTAL	
	n_i	m_i (g)	n_i	m_i (g)	n_i	m_i (g)	n_i	m_i (g)	N_i	M_i (g)
Porifera										
Haliclona			1	10					1	10.0
Nemertea										
Lineus	1	0.1			3	0.3	1	0.1	5	0.5
Annelida Polychaeta										
Glycera							3	3.5	3	3.5
Nereis							1	2.7	1	2.7
Diapatra							5	12.5	5	12.5
Crustacea										
Balanus	28	5.5			7	1.3			35	6.8
Gammarus	50	5.0	16	1.8	12	1.5			78	8.3
Mollusca Gastropoda										
Acmaea	2	1.3							2	1.3
Littorina	40	78.0	7	13.7	17	33.6	1	0.4	65	125.7
Nassarius	4	0.3	89	67.0	19	15.0			112	82.3
Thais	6	12.0			5	10.5			11	22.5
Bivalvia										
Macoma							2	30.3	2	30.3
Mya	1	25.0			1	33.0	16	48.0	18	106.0
Mytilus	27	113.0	80	365.0					107	478.0
Echinodermata Asteroidea										
Asterias	1	15.0			5	79.0			6	94.0
Echinoidea										
Arbacia	3	51.0	2	99.0	1	23.0			6	178.0
Total =	N_s	M_s	N_s	M_s	N_s	M_s	N_s	M_s	N_t	M_t
(station) =	163	306.2	201	556.5	70	197.2	29	97.5	457	1162.4
Species (Spp) =	11		6		9		7		16	

A2.2). The three parameters (NC, GC, FO) are computed for each species (using the data from Table A2.2) and used to calculate IRI:

IRI = (NC + GC) × FO

Because each parameter can range from 0 to 100%, the minimum IRI = 0; the maximum would be IRI = 20,000 for a situation in which only one species existed in all the stations examined.

The full array of values for the sample data is presented in Table A2.5. As another example, sample calculations are provided for *Mytilus* as a footnote at the bottom of the table. Note that the species are ranked according to IRI.

Graphical displays of n_i and FO (Fig. A2.4), NC and GC (Fig. A2.5), for the six highest-ranked species (based on IRI) are also provided. You should compare the relevant information conveyed in these illustrations.

Table A2.3 Calculation of diversity indices (D) and (H) for macrofaunal invertebrate survey, Lamoine Pool station. Data from Table A2.2.

Species	n_i	$(n_i/N_s)^2$	$\ln (n_i/N_s)$	$(n_i/N_s) \ln (n_i/N_s)$
Lineus	1	3.76e−5	−5.094	−0.031
Balanus	28	2.95e−2	−1.762	−0.303
Gammarus	50	9.41e−2	−1.181	−0.362
Acmaea	2	1.15e−4	−4.401	−0.054
Littorina	40	6.02e−2	−1.405	−0.345
Nassarius	4	6.02e−4	−3.707	−0.091
Thais	6	1.36e−3	−3.302	−0.122
Mya	1	3.76e−5	−5.094	−0.031
Mytilus	27	2.74e−2	−1.798	−0.298
Asterias	1	3.76e−5	−5.094	−0.031
Arbacia	3	3.39e−4	−3.995	−0.074

Total (= N_s) = 163   0.214   −1.742

$D = 1 - \Sigma (n_i/N_s)^2$ $H = 1 - \Sigma (n_i/N_s) \ln (n_i/N_s)$

$D = 1 - 0.214$; $D = 0.786$ $H = -(-1.742)$; $H = 1.742$

Using the same procedure for the other stations:

Lamoine Jetty $D = 0.638$ $H = 1.118$
    Haliclona
    Gammarus
    Littorina
    Nassarius
    Mytilus
    Arbacia

Sandy Point $D = 0.826$ $H = 1.789$
    Lineus
    Balanus
    Gammarus
    Littorina
    Nassarius
    Thais
    Mya
    Asterias
    Arbacia

Fort Point $D = 0.646$ $H = 1.399$
    Lineus
    Glycera
    Nereis
    Diapatra
    Littorina
    Macoma
    Mya

Table A2.4 Similarity index (SI) computed for every combination of the four stations sampled for macrofaunal invertebrates (see Table A2.2).

Station	Lamoine Pool	Lamoine Jetty	Sandy Point	Fort Point
Lamoine Pool	—			
Lamoine Jetty	0.59	—		
Sandy Point	0.90	0.53	—	
Fort Point	0.33	0.15	0.38	—

SI = 2C/A + B
SI = 2 × (number of species common to both stations)

(number of species at one station + number from second)

For Lamoine Pool and Lamoine Jetty:

SI = (2 × 5)/(11 + 6) = 0.588

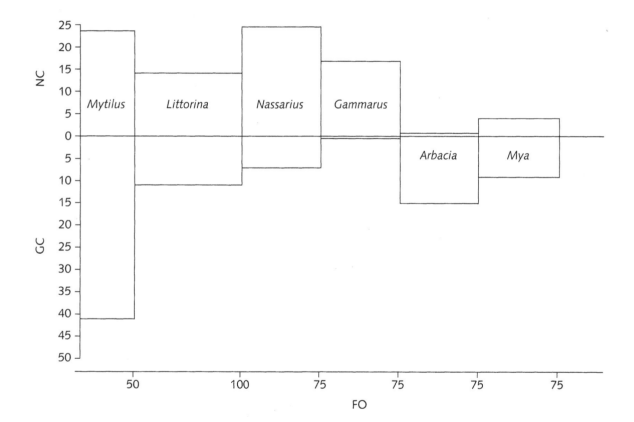

Figure A2.3 IRI diagram of the six highest-ranked species in the field samples listed in Table A2.2. Each species is represented by a rectangle, with the vertical axis consisting of numerical composition (NC) above the horizontal baseline and gravimetric composition (GC) below it. The horizontal axis represents frequency of occurrence (FO). Note that the horizontal axis is located at zero for each species. Computations of NC, GC, and FO (shown in Table A2.5) are based on field data from Table A2.2. The IRI essentially represents an area: the greater the IRI, the larger the rectangle. A species represented by a few, but large individuals has a diagram similar to that of *Arbacia;* one with a large number of small individuals has a diagram similar to that of *Gammarus. Mytilus* is an example of a species represented by nearly equal numbers and biomass (weight). Compare this figure with Figures A2.4 and A2.5, in which the same data are plotted independently.

Table A2.5 Index of relative importance calculations of intertidal invertebrates (16 species) from four stations (= St) on Maine coast (see Table A2.2).

Taxa	Total			NC	GC	FO		
	N_i	M_i (g)	S_o	%N_t	%M_t	%St	IRI	Rank
Porifera								
Haliclona	1	10	1	0.23	0.86	25	27.25	13
Nemertea								
Lineus	5	0.5	3	1.09	0.04	75	84.75	10
Annelida Polychaeta								
Glycera	3	3.5	1	0.66	0.31	25	24.00	14
Nereis	1	2.7	1	0.23	0.23	25	11.50	16
Diapatra	5	12.5	1	1.09	1.07	25	54.00	12
Crustacea								
Balanus	35	6.8	2	7.66	0.59	50	412.50	8
Gammarus	78	8.3	3	17.07	0.71	75	1333.50	4
Mollusca Gastropoda								
Acmaea	2	1.3	1	0.44	0.11	25	13.75	15
Littorina	65	125.7	4	14.22	10.81	100	2503.00	2
Nassarius	112	82.3	3	24.50	7.08	75	2368.50	3
Thais	11	22.5	2	2.41	1.94	50	217.50	9
Bivalvia								
Macoma	2	30.3	1	0.41	2.61	25	76.25	11
Mya	18	106.0	3	3.93	9.12	75	978.75	6
Mytilus*	107	478.0	2	23.41	41.12	50	3226.50	1
Echinodermata Asteroidea								
Asterias	6	94.0	2	1.31	8.09	50	470.00	7
Echinoidea								
Arbacia	6	178.0	3	1.31	15.31	75	1246.50	5
	N_t	M_t						
Totals	457	1162.4		100.0	100.0		1304.25	

*Sample Calculations for *Mytilus*:
NC = [107 individuals (= N_i)/457 individuals (N_t)] × 100 = 23.41%
FO = *Mytilus* found at 2 (= S_o) of 4 stations (= St); FO = 2/4 = 50%
GC = [478 g (= M_i)/1162.4 g (= M_t)] × 100 = 41.12%
IRI = (NC + GC) × FO = (23.41 + 41.12) × 50 = 3226.5

DATA PRESENTATION

Tables

The **table** presents data in a visually organized, yet still exact format. It contains large amounts of information in a form (accurate numbers) that is readily accessible to the reader. Each table (see Table A2.1) contains a title (placed at the top) and three horizontal lines that separate the title, column headings, data field, and footnotes. Examples of other tables are found throughout the manual. Tables are not considered figures or illustrations.

Graphing Data

Graphs are used to show amounts, frequencies, trends, or relationships of data. The type of graph used depends on the purpose of the graph and the type of data. Follow these basic guidelines to produce acceptable graphs, and then select the type of graph that is most appropriate for your data set.

1. Use a *ruler;* never draw a line free-hand.
2. Place the axes and their labels to maintain a *one-inch margin* on all four sides of the page.
3. Set the *tick marks* along the axes so that they extend outward and are equally spaced.

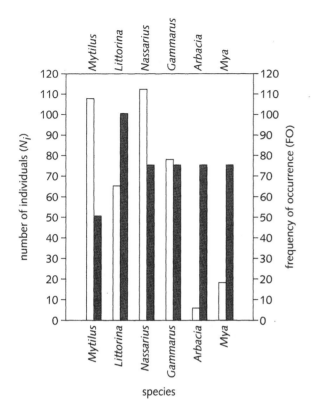

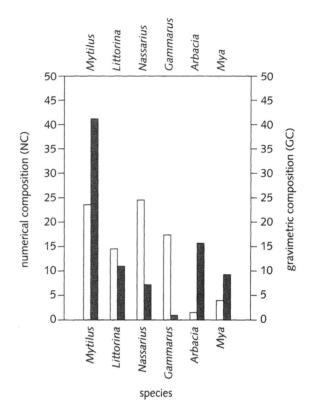

Figure A2.4 Bar graph showing two parameters (number of individuals = N_i and frequency of occurrence = FO) for six species collected from four stations along the Maine coast (see Table A2.2). The open bar for each species represents total N_i (from Tables A2.2 and A2.5); the hatched bar is the FO (from Table A2.5). The species are listed in order of rank based upon their index of relative importance (IRI) (Table A2.5 and Figure A2.3). While it is possible to plot all 16 species in this fashion, only six were selected to simplify the demonstration of this graph. Note that the abundance (number of individuals) of a species is not indicative of its FO. Many individuals of a species might occur at only one station, resulting in a high n_i value and a low FO. In a situation where an identical n_i is distributed over all four stations, the FO would be high. Note that the labels of the y axes have the same orientation.

Figure A2.5 Bar graph of numerical and gravimetric composition (NC and GC) for the six highest-ranked species, based upon IRI values (see Table A2.5 and Figure A2.3). Note that NC and GC cannot be used to predict each other. One large animal results in a very low NC and a high GC. Conversely, numerous small animals are indicated by a high NC with a low GC.

4. *Label* the tick marks and each *axis* to indicate clearly the *parameter* and its *units of measurement*.
5. Connect the dots in a *line graph*. Do not visually fit a smoothed line to the data points; fitting is only appropriate if done mathematically.
6. For a *scatterplot*, no line is necessary if the trend or relationship is conspicuous. If you need to add a line, calculate its position by regression analysis.

7. Write a *caption* or *legend* for your figure. In combination with your graph, this description should thoroughly reveal to the reader what occurred and the result. Captions of figures are placed below the illustration or on a separate page. Ask your instructor for advice.

Line Graphs

Line graphs are used to show a relationship between variables of *continuous data*, such as the trend over time, or the dependence of a response on the concentration of a stimulant. Such variables have constant intervals between successive units and a zero point that has a physical meaning—for example, weight, volume, or time (Fig. A2.2). Continuous variables are plotted as a line because we perceive the continuous line to represent the continuous form of the data. A line graph has two axes, at right angles to each other, that represent the scales of two continuous data variables. The axes form a coordinate grid, on which a relationship between the two variables is shown. One variable may be *dependent* on the other; the value of the dependent variable changes as the independent variable changes. An *independent* variable is one for which the investigator sets a value or range of values; it is often controlled in an experiment to observe the response of the dependent variable.

Conventionally, the independent variable is graphed horizontally, along the *x axis (abscissa),* and the dependent variable is graphed vertically, along the *y axis (ordinate).*

In many field studies, you may not be able to change either variable and may not recognize which one controls the other. In that case, the dependent variable is selected arbitrarily. Such data may result in a line graph, but if your data set consists of multiple pairs of measurements taken over a range of values, your data might better fit into a scatterplot.

Scatterplots

A **scatterplot** is used to show whether a correlation exists between two continuous data variables (Fig. A2.1). This graph is drawn on a coordinate grid on which a single (datum) point represents each event or value. Whether the variables shown on the axes are correlated and to what degree can be determined by trying to fit a mathematical function to the data points (see the section on regression analysis). This action permits you to fit a smooth curve that may not pass through every data point.

Plotting Several Variables on the Same Graph

It is often instructive to display two or more variables on the same plot, such as in Figure A2.2. Follow the guidelines below and those listed in the section on graphing data.

1. Provide each variable with its own axis scale, clearly labeled to avoid confusion with other variables.

2. Plot each variable with its own distinctive set of data points and lines. Acceptable symbols are open and closed circles, triangles, and squares. Acceptable lines are solid, dashed, and dots. Do not use colors to distinguish between data sets; they all photocopy black.

3. If the separate lines on the graph are too difficult to follow, make several smaller graphs on one sheet of graph paper. If graphs are arranged so that their common variable is in alignment, comparisons will be easier.

Bar Graphs

Data collected according to categories described by name (nominal = discrete data) can be displayed in bar graphs, in which each of the bars (rectangles) along one axis is identified by a title (Figs. A2.4 and A2.5). The other axis is divided into units of measurement. Because the information for any single category is unrelated (discontinuous) to any of the other categories, data categories should never be connected to each other by a line. The order of categories on the bar graph is somewhat arbitrary, but concern should be given to some logical arrangement, such as taxonomic, chronologic, or sequential. A bar graph may sometimes be used with continuous data to depict frequency distributions.

Frequency Distributions

Studies of populations of organisms often include recording weight, length, volume, or other measurement of individuals sampled from a population. You may find it helpful to describe the population composition (see the section on descriptive statistics). These data can be subdivided into size classes (body dimensions, weight, volume) in tabular form and then plotted as a histogram. Such plots clearly indicate how your data are distributed: *normal* (Fig. A2.6); *skewed* (Fig. A2.7); *bimodal* (Fig. A2.8); or *uniform* (Fig. A2.9). Compare the distribution of your population with the mathematics of central tendency (mean, median, mode, standard deviation). Follow these guidelines for the design of a histogram.

1. Consider using a histogram only if you have data for more than 20 individuals.
2. Divide your data into equal class intervals, and consider the following:

    a. Select the class intervals so no overlap occurs (for example, 1.0–1.9; 2.0–2.9).

    b. Divide the data into the number of classes that will allow you to determine the overall pattern. As a rule, you should not have fewer than 5, nor more than 15, classes.

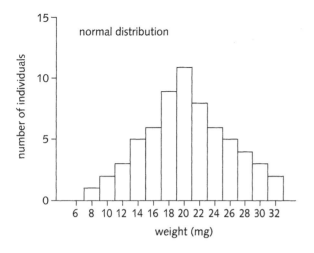

Figure A2.6 Sample histogram of a normal size frequency distribution of a single species. Size, in this case, is weight of the organism. The data were arbitrarily arranged into 2-mg size classes along the *x* axis. This particular distribution is considered "normal" because it contains nearly an equal number of individuals on both sides of the mode (20 mg), and the progression (increase and decrease in value from the mode) is regular and gradual (normal). The arithmetic mean ($\overline{X}$) is very close to the mode in this case, and nearly all of the values will fall within two standard deviations (SD) of the mean.

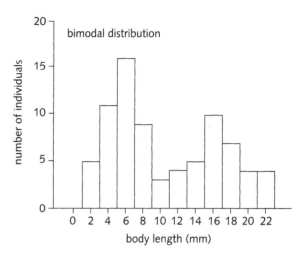

Figure A2.8 A bimodal frequency distribution is revealed by two clearly separated modes (6 mm and 16 mm). A bimodal distribution can be considered to be two normal distributions that overlap. In this example, the two modes could represent two year-classes. The younger (smaller) distribution may contain a few individuals of 10 mm, and even 12 mm in length, while the older (larger) distribution could include some relatively small (12 mm or 10 mm) individuals.

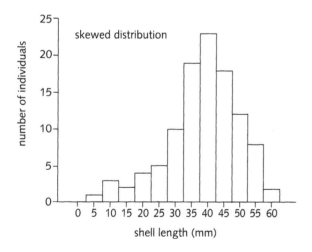

Figure A2.7 A skewed size frequency distribution based on shell length arrayed into 5-mm size classes. The mode (40 mm) appears shifted (skewed) to the right. The distribution is represented by a larger size range on the left (smaller shell length) than on the right of the mode.

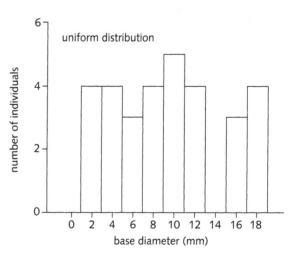

Figure A2.9 A uniform frequency distribution exists when no mode is evident. All size classes are represented by nearly equal numbers of specimens.

3. Draw the rectangles (use a ruler) over the class intervals to a height corresponding to the frequency scale on the *y* axis. The axis label must indicate whether you are plotting counts or percentages of the population.

Oceanographic Data in Relation to Depth

Field observations are made at different depths. The standard method is to place **depth** on the vertical axis with surface (= 0) at the top and increasingly greater depths below (Fig. A2.10). The dependent variable is placed on the horizontal axis at the top, with zero or low values to the left.

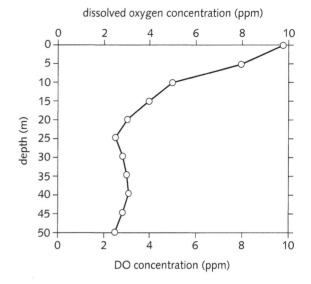

Figure A2.10 Parameters measured at varying depths (of water or sediment, for example) are best illustrated by placing the intersection of the *y* and *x* axes in the upper left corner of the graph. The values of *y* and *x* then increase downward and to the right, respectively. This line graph shows that eleven measurements of DO were made at 5-m depth intervals. The data are plotted as open circles connected by a solid line. The graph reveals a high DO concentration at the surface, a rapidly diminishing oxygen content to a depth of 10 m, a more gradual decrease to 25 m, and a somewhat variable, but low concentration in deeper water. More than one variable can be plotted on such a graph if different symbols and lines are used for each parameter, and the data do not overlap or mask each other.

APPENDIX 3

Writing Scientific Reports

The laboratory portion of this course provides you with experience in acquiring and interpreting data. Data acquisition is one of the enjoyable parts of research. Interpreting the data and observations brings the laboratory or field experience to a meaningful conclusion.

You can interpret data in a simplified format by answering a series of leading questions in laboratory report forms, a method used in several units in this manual. Other studies, such as field trips, require a more sophisticated approach to data interpretation—the written report.

Learning to write effective laboratory reports represents an investment in your future. Preparation of reports develops the ability to think clearly, organize ideas logically, and express yourself accurately and concisely. Such skills are assets in any career.

This appendix provides you with an overview of a written scientific report and an outline of the fundamental principles of effective technical communication. For more detailed information, consult a technical writing text, such as MacMillan (1988), Day (1994), Pechenik (1993), or Day (1995). The first text includes a sample research paper that may be used as a model.

COMPONENTS OF THE LABORATORY OR FIELD TRIP REPORT

1. Include a *cover page* that includes the *title* of the project or study (name of the exercise), your *name* and program, course name, and *date* of submission.
2. The *introduction*, usually one or two paragraphs in length, presents the question(s) asked and the objective and purpose of the study. It also contains a brief summary of relevant background information. If a hypothesis is tested, it should be stated in one sentence, near the end of this section.

3. In the *materials and methods* section you describe what you did and how you did it. Write chronologically, in past tense. Include all your methods in sufficient detail to allow others to repeat your study if they desire.
4. The *results* section contains your data and a narrative of your observations. Present tables and figures as your evidence. Confine each table and figure to a single page, and keep those pages separate from the narrative text. The text summarizes your findings clearly and objectively, without bias or interpretation. The design of *tables* and *figures* is described in more detail in Appendix 2 in this manual.
5. The *discussion* section allows you to *interpret* your observations. Compare and contrast your data to information you obtained from background reading. Evaluate your results in terms of the original hypotheses or questions asked. Explain the biological significance of your results—that is, draw *conclusions*. Try to formulate new hypotheses based upon your observations. Report any statistical tests applied to your data. If your results are unexpected or contradictory, attempt to *explain* the discrepancies, and suggest pathways for further study.
6. Include the full *citations* of any *references* (textbooks, laboratory handouts) you mentioned in the *literature cited* section. Cite only material that you read or used in the paper. Some examples of style follow:

For a Research Paper

Cain, A. E. 1993. The life cycle of *Mnemiopsis leidyi*. Journal of Natural History 17:331–342 .

Abel, B. C., A. E. Cain and E. Adam. 1992. Variations in the life cycle of *Mnemiopsis leidyi*. Bulletin of Marine Science 33:222–233.

For a Book (Monograph)

Haefner, P. A., Jr. 1996. Exploring Marine Biology: Laboratory and Field Exercises. D. C. Heath and Company, Lexington, MA, 272 pp.

For a Chapter in an Edited Monograph

Crabbe, B. A. 1933. Hormonal control of ecdysis. *In:* R. U. A. Shrimp (ed.), The biology of Crustacea, Vol. 13, Academic Press, New York, pp. 26–47.

BASIC PRINCIPLES OF WRITING

The credibility of your writing can be damaged by basic grammatical errors, such as illogical, fragmented, or run-on sentences, faulty punctuation, or poorly chosen words. Such errors confuse and annoy the reader. In this section you will find **some basic rules of effective writing** that the casual writer commonly violates. You will also find conventions that apply to **reporting and writing numerals, use of scientific names of organisms** in reports, and the problem with **teleology.** Apply the rules to your writing.

Basic Rules of Effective Writing

1. Write complete, simple declarative sentences.

   This is a simple declarative sentence.

2. Subject and verb must agree.

   Each of the nets **was** damaged.

3. A pronoun must agree in number with its antecedent.

   The fish avoided the **crab and worm** because **they** were dead.

4. Do not shift verb tenses unnecessarily. Be consistent.

   She **dispensed** the reagent, and **completed** the assay.

5. Place modifiers as close as possible to words modified.

   The **crab** is not eating, and **it** has not been active.

6. Make comparisons complete.

   The 35 o/oo sample is saltier **than the 30 o/oo sample.**

7. Use superlatives and diminutives correctly.

   The hard clam yielded **less** protein than the soft clam. Of the three bivalve extracts analyzed, the hard clam extract contained the **least** amount of protein.

Reporting Numbers

1. In reporting a decimal number, retain no more digits to the right of the decimal point than the precision warranted by your experimental method.

2. To "round" an inconveniently large number, simplify it:

   2,645,381 can be expressed as 2.6 million.

3. When reporting data in which the values are less than one, always *include the zero* to the left of the decimal point:

   0.1 psi   0.065 mm   0.999 km

Writing Numbers

1. Use a numeral or numerals:

   a. To express any number that immediately precedes a standard unit of measure or its abbreviation:

      1 gram   3 g   18 mm

   b. For a date, an expression of time, a page number, a percentage, a decimal quantity, or a numerical designation:

      7 January 197   7 Jan 71   page 1079
      27 percent   27%   37.6 g
      the time is 0815   a magnification of 50

   c. For a number implying arithmetical manipulation:

      18 multiplied by 2   a factor of 2

   d. For numbers grouped for comparison or having statistical implications:

      Sizes ranged from 9 mm to 54 mm.
      Collections 2, 4, 13, and 27 have been studied.

2. In most situations, use words for numbers one through nine, and numerals for larger numbers:

   nine rabbits   two dogs   14 parts

3. In a series containing some numbers of 10 or more and some less than 10, use numerals for all:

   the 7 apple trees, 9 peach trees, and 20 plum trees

4. Treat ordinal numbers as you would cardinal numbers:

   third   fourth   33rd

5. In writing a large number ending in several zeros, either substitute a word for part of the number or add an appropriate prefix to a basic unit of measurement:

1.6 million (*not* 1,600,000)
23 μg (*not* 0.000023 g)

6. Do not begin a sentence with a numeral; either spell out the numeral, reword the sentence, or end the preceding sentence with a semicolon:

Six hundred students peeled 600 potatoes.
The **600** potatoes were peeled by 600 students.

7. If one number immediately precedes another, spell out the first and use a numeral to express the second:

    fourteen 10-cm rats   thirty 5-foot beams

Use of Scientific Names

1. **Common names** of plants and animals are derived from traditions of folklore and popular usage. Generally, common names are misleading and do not find universal acceptance.

   It is not good practice to refer to an organism solely by its common name. Refer to organisms by common names only if you have first given full scientific names and other taxonomic information.

   The grey trout, *Cynoscion regalis* . . .

2. Each species has just one **scientific name.** The name is **binominal** because it contains two parts: the *genus* to which the organism belongs, and the particular *species*.

   a. The species name of the organism is always expressed by the two words. For example, the genus of the rock crab is *Cancer*; the species is *Cancer irroratus*.

   b. In (a), the scientific name, or species, is printed in italicized type. When the printer is unable to print italics, use underlining. This rule also applies to hand-written or hand-printed documents.

   c. Note that the genus is always capitalized; the second (species) name is not.

3. The scientist who first publishes the description of the newly recognized species is considered its "author."

   a. The author's last name is placed after the species name: *Callinectes sapidus* Rathbun. Some authors' names are abbreviated: R. for

Rafinesque; L. for Linnaeus; Fabr. for Fabricius.

   b. An author's name is placed in parentheses to indicate that the species has been placed into a different genus from the original: *Kickxia spuria* (L.).

   c. Give the full scientific name (genus and species) the first time a species is mentioned in the text. In later references to this species, you may abbreviate the genus by its first letter (still capitalized and italicized). *Cancer irroratus* becomes *C. irroratus*. Note that confusion may arise if you also refer to *C. sapidus* in the same paragraph. In such a case, write out both full names.

4. Although it is common in textbooks to see the term *Nereis* or *Metridium*, such a reference is incorrect.

   a. The statement should refer to "a species of *Nereis*," or to "*Nereis* sp." The "sp." indicates that the species is unknown.

   b. Note that "sp." is not italicized. If more than one species is referenced in this fashion, the abbreviation is "spp."

5. Do not put an article (the, a, an) immediately before the scientific name—for example, "The *Callinectes sapidus* is the common blue crab." It is correct to state, "*Callinectes sapidus* is the common blue crab."

6. Scientific names are never written as plurals, as "Blue crabs, *Callinectes sapidi*, were observed on the beach." The correct version is "Blue crabs, *Callinectes sapidus*, were observed on the beach."

7. Except in keys (guides to identification) or other taxonomic writings, the specific name of an organism must always be preceded by the generic name or its abbreviation.

   a. It is incorrect to state, "*Callinectes* was the most common crab on the beach." It is correct to write, "*Callinectes sapidus* was the most common crab on the beach."

   b. It is not correct to state, "*C. sapidus* was the . . ." Never begin a sentence with an abbreviation.

8. Names of genera may be used alone if you are referring collectively to the species in a particular genus. For example: "Some species of *Sargassum* grow in dense mats on the surface of the ocean."

9. Taxonomic groups (taxa) above the level of genus (family, order, class, phylum, and so on) are capitalized but not italicized. For example, you would write: "The Bivalvia (clams and oysters) and the Gastropoda (snails) are two of the classes in the phylum Mollusca."

10. Biologists sometimes drop or modify the endings of taxa to make common names—for example, gastropods from Gastropoda; polychaetes from Polychaeta.

11. Common names are not normally capitalized except in accordance with specific guidelines for certain groups. If a common name is derived from a particular person or place, that word is capitalized even if the rest of the common name is not: American lobster, European eel.

Teleology

Teleology has at least three definitions:

1. The fact or the character of being directed toward an end or shaped by a purpose.
2. The doctrine or belief that design is apparent or ends are imminent, in nature; especially, the vitalist doctrine that the processes of life are not exclusively determined by mechanical causes, but are directed to the realization of certain normal wholes.
3. Attributing a sense of purpose to other living things, especially when discussing evolution.

Giraffes did not evolve long necks "in order to reach the leaves of tall trees." Insects did not evolve wings "in order to fly." Organisms cannot evolve structures, physiological adaptations, or behavior out of desire. Appropriate genetic combinations always arise by random genetic events, by chance, before selection can operate. Selection is then imposed on the individual by its surroundings. In that sense, selection is a passive process. Natural selection never involves conscious, deliberate choice, and it is incorrect to write statements that imply otherwise.

Note that many teleological statements contain the dangerous combination of words "in order to." When you describe a biological process, **avoid** the terms "in order to." For example, "The crab molts in order to grow." This statement implies that the crab thought about the process and then did something about it. A crab cannot make a decision to grow. It would be better to state, "Growth in crabs is made possible by periodic shedding of the shell, a process known as molting."

Avoid phrases such as "so that it can," "which allows it to," and "which enables it to." They are dangerous, and lead only to trouble.

Study these teleological examples:

1. Incorrect: Parent gulls remove white, conspicuous eggs to protect the newly hatched, black-headed young.
   Correct: Parental removal of white, conspicuous eggs may protect the newly hatched, black-headed young from predation.
2. Incorrect: Male fiddler crabs have one enlarged cheliped in order to attract females.
   Correct: Male fiddler crabs use the enlarged cheliped in a mating display.
3. Incorrect: Macroscopic algae have holdfasts to anchor themselves to firm substrates.
   Correct: Macroscopic algae are anchored to firm substrates by holdfasts.

You should never be prompted to write teleologically. Stay constantly on the alert for questions (in texts, quizzes, and reports) that are worded like this example:

Why does this organism have this structure?

Ask for clarification whenever you encounter such a question. It would be better worded as:

Of what adaptive advantage is this structure to the organism?
What is the function of this structure?, or
What does this structure enable the organism to do?

REFERENCES

Day, R. A. 1994. How to write and publish a scientific paper. Oryx Press, Phoenix, Arizona, 223 pp.

Day, R. A. 1995. Scientific English, a guide for scientists and other professionals. Oryx Press, Phoenix, Arizona, 148 pp.

MacMillan, V. E. 1988. Writing papers in the biological sciences. St. Martin's Press, New York, 142 pp.

Pechenik, J. A. 1993. A short guide to writing about biology. Harper Collins Publ., New York

CREDITS

Text

p. 23, Invertebrate Key has been highly modified from K. L. Gosner, *Guide to Identification of Marine and Estuarine Invertebrates, Cape Hatteras to the Bay of Fundy*, Wiley-Interscience, NY. Reprinted by permission of Mrs. Kenneth W. Gosner.

p. 126, Plankton Key modified from D. L. Smith, *A Guide to Marine Coastal Plankton and Marine Invertebrate Larvae.* Copyright 1977 by Kendall/Hunt Publishing Co. Used with permission.

p. 150, Fish Key slightly modified from E. O. Murdy, *Saltwater Fishes of Texas, a Dichotomous Key.* © 1983 Texas A&M University Sea Grant College Program. Reprinted by permission.

p. 199, Table 12.4, "Abundance Scales for some rocky shore organisms" modified from S. J. Hawkins and H. D. Jones, 1992, *Marine Field Course Guide, Rocky Shores.* Reprinted by permission of Immel Publishers, Berkeley Square, London.

Photos

p. 10, Aquatic Eco-Systems, Inc.; p. 38 (top), Andrew Martinez/Photo Researchers; p. 38 (middle), Paul A. Haefner, Jr.; p. 39, Andrew J. Martinez/Photo Researchers; p. 45, Mark Newman/Photo Researchers; p. 118 (left and right), Wildlife Supply Company; p. 175, Eunice Harris/Photo Researchers; pp. 218 and 223, Wildlife Supply Company; pp. A2 and A4 (left and right), Leica Inc.

Printed in Australia
AUHW011229010622
364440AU00005B/17